HOME BREWER'S GOLD

Other Avon Books by
Charlie Papazian

THE HOME BREWER'S COMPANION
THE NEW COMPLETE JOY OF HOME BREWING

HOME BREWER'S GOLD

Prize-Winning Recipes from the 1996 World Beer Cup Competition

CHARLIE PAPAZIAN

AVON BOOKS ◆ NEW YORK

Home Brewer's Gold is published in cooperation with the World Beer Cup and Brewers Publications, both of which are departments of the Association of Brewers, P.O. Box 1679, Boulder, CO 80306 USA (303) 447–0816.

AVON BOOKS
A division of
The Hearst Corporation
1350 Avenue of the Americas
New York, New York 10019

Library of Congress Cataloging in Publication Data

Papazian, Charlie.
  Home brewer's gold / Charlie Papazian.
    p. cm
  Includes index.
  ISBN 0-380-79192-7
  1. Brewing—Amateurs' manuals. I. Title.
TP570.P284 1997   97-23662
641.8'73—DC21   CIP

First Avon Books Trade Printing: December 1997

AVON TRADEMARK REG. U.S. PAT. OFF. AND IN OTHER COUNTRIES, MARCA REGISTRADA, HECHO EN U.S.A.

Printed in the U.S.A.

QPM   10 9 8 7 6 5 4 3 2 1

Contents

Contents

Preface

Home Brewer's Gold is designed and written with a respect for the devotion of the world's commercial brewers and the passion of today's homebrewers. Its intention is to help inspire the evolution of today's beer cultures and to develop even greater appreciation for the craft of brewing.

This book is about more than just homebrewing the winning beers of the 1996 World Beer Cup International Competition. The origins of every brewery can be traced to the spirit of home. Whether it was simple appreciation with friends and family or a passion to brew, one can certainly find an essential flame, an essential spirit belonging to all brewers.

Homebrewers are special. In every century, in every land, in every community, homebrewers represent the appreciation of craft, art and science and the special way brewing can contribute to the quality life in all communities. Today's homebrewers find themselves admired for their creativity and their passion for revitalizing an appreciation for the world's great beers, past, present and future.

I could not have developed, formulated and written this book without the participation of all brewers. Winning brewers were most generous in providing brewery and beer recipe information that helped in my recipe formulations.

Special thank-you's to Marcia Schirmer, who directed the event; to Jeanne and Glenn Colon-Bonet, who managed the competition; to Sharon Mowry, who managed all of the shipping and organizing of the beer; and to Rebecca Bradford, who managed the staging. Thanks also go to Sheri Winter, Cathy Ewing, Bob Pease, Kyle Keazer and Melinda Bywaters, all staff of the Association of Brewers who contributed to the production of the event. Special acknowledgment must be given to the U.S. Bureau of Alcohol, Tobacco and Firearms (BATF). Their understanding of the World Beer Cup purpose and their cooperation and assistance in helping entries legally clear U.S. Customs made the event possible in the United States. A big thank-you to Mark Snyder, my project assistant, who collected all of the brewery-related data and information. Again, as with my book *The Home Brewer's Companion*, I thank Tracy Loysen for helping me with preliminary editing of my manuscript. Finally, thank-you to Tom Nelson of TKO Software, developer of "Brewers Workshop," used to assist me in calculating recipes.

Introduction:
The Beer Tree

It was a brisk, chilly winter's afternoon in midtown New York City. I had been in the city and Long Island visiting local brewpubs and homebrewers, and on this particular afternoon my editor, Stephen Power at Avon Books, had invited me to drop by to see him and have lunch at a small Italian restaurant. He had an idea he wanted to pass by me. The traffic outside and the waiters and waitresses inside scurried around, relatively oblivious to the quiet conversations at each table. "I'll have a ——— beer, please." "I'll have the same," added Stephen. What beer did we choose? Well, if the brewer ever asks, I'll confess, but that will be between Stephen, the brewer and me, for the time being.

Now, what was this idea? "I came across the Wall Street Journal's article about your Association of Brewers' announcement of The World Beer Cup. . . . What about a book on brewing the winning recipes?" That was halfway through the first beer.

My immediate reactions were two. Why didn't I think of that? and How the hell am I going to be able to formulate nearly sixty recipes and maintain my credibility? I finished my beer and had another. Things became clearer.

It's not my full-time job to think of ideas for books. The first question was easy to answer. Stephen was good at his job. Now then, could I do it? After all, I truly admire and respect the brewers around the world and the quality of their products. The idea of attempting to formulate and present 5-gallon recipes that match the character of their beer was, well, intimidating. I had always invented my own recipes before. I'd hardly ever tried to duplicate someone else's beer. The challenge intrigued me. I thought of having another beer, but I knew better. So I did. "Let me think about this one, Stephen. I know it could be done. It's whether I can make a quality presentation and feel I've ethically represented both the World Beer Cup and the winning beers and breweries in a fair fashion, while offering practical and useful information to homebrewers." Those weren't my exact words, but surely these thoughts whirled in my head at such a grandly bold project.

One month later I decided. Yes I would. What remained so uncertain was whether I could really formulate recipes from just tasting the beer. This was the bottom line. There was a possibility I would have to rely solely on my own skills as a brewer and taster. Yes, I would ask each of the winning breweries for technical information regarding ingredients and measurements, but how forthcoming would they be? What if none of them was willing to divulge information, considering it a trade secret? Then I thought, "Hell, I've been brewing for twenty-six years and that's got to be worth something."

The World Beer Cup was staged in June of 1996. The winners were announced and my work began. But I really didn't realize the magnitude of what I had gotten myself into until the van delivered to my home fifteen cases of beer, two bottles each of the first-, second- and third-place winning beers. They were cold when they arrived and were immediately placed into a 6-by-9-foot walk-in cooler that I had just finished having constructed especially for these beers. The fifteen cases of beer secured, I slammed the cooler door shut. As the delivery van pulled away that warm Tuesday summer afternoon, I sat on a wooden crate. I can't quite sort out exactly what I was feeling. I don't think I ever will. Was it total elation, joy, and whoopee? I was going to have to drink 146 different kinds of the best beers in the world at my leisure at home. Or was it an overwhelming feeling of despair and hopelessness overcoming me? I had to taste and take notes on 146 different beers and formulate recipes for 52 of them.

I was in a quandary, so I poured myself a bottle of home-brew, looked at it, smelled it, drank the whole thing, and then resigned myself with a sigh. "Oh well, it's a tough job, but someone's got to do it."

And this is how the summer of 1996 officially got under way for me and my favorite tree. You see, I was thinking of each and every one of you who are homebrewers, beer enthusiasts and professional brewers every time I poured myself one of those beers. You would have often found me outside at the picnic table on my deck, tasting four to eight beers at a sitting, laptop computer positioned front and center, emitting its eerie light onto my face, my special tasting glass always within arm's reach, as well as a pitcher of water for rinsing glasses and the Styrofoam box maintaining the beer's chill. I was thinking of you every time I finished evaluating one of those 146 different and remarkably great beers. The pleasure was all mine, but only two or three ounces at a time. Oh yes, I thought of all of you each time the remainders went shooting into the air, cast over my shoulder, splashing against what has become for me over time a very special tree. The tree's roots awaited the agitated white and brown beer foam that trickled down its trunk and into the ground. That summer it was a very happy tree.

But now it's your turn to experience what my friend the beer tree enjoyed all summer long. Using your skills as a homebrewer in conjunction with my recipe formulations and tasting notes, I believe that you, too, can enjoy the authentic character of some of the world's best beers. The most important thought to keep in mind is that a recipe does not a beer make. A beer is made using the skills of a brewer and the interaction of ingredients, process and packaging—

all unique circumstances for every brewer in the world. The influence that equipment, process control and particular ingredients have on the final product cannot be underestimated. All of us as brewers have developed our own techniques and equipment adaptations. In many cases these recipes will only be a beginning point for achieving the winning character. For many of you the first attempt at using these recipes will be an unparalleled success. Surely each time you will brew excellent beer. But getting the exact balance of characters may take time as you adjust these recipes to conform to your own manner of brewing. That is what the craft of homebrewing is all about—knowing when and how to adapt to achieve the desired end.

Without any doubt in my mind, the most important thought I wish to convey to each and every one of you is the admiration I have for all of the brewers who entered the World Beer Cup. The entries were statements recognizing their own pride and the pride they have in the international brewing community. Theirs was a gesture stating, "I think my beer is good enough to represent the brewing community and is worthy of being recognized as such." When a beer and a brewery win an award in the World Beer Cup, they then represent the quality all the world's brewers strive for. They become ambassadors for all the world's porters, Pilseners, bocks, brown ales . . . all lagers and all ales. It is the brewing community's quality we try to replicate as homebrewers. And in turn, the professionals often try to replicate the quality of our homebrews.

—Charlie Papazian
Boulder, Colorado, USA

The Association of Brewers World Beer Cup International Beer Competition

Beer is a global phenomenon. It always was and always will be. As our world continues to change and evolve, so do the marketplaces in which beer is presented. In many countries new breweries' beers are commonly presented next to recognized beer classics.

The quality and diversity of beers available from around the world prove that brewing excellence knows no borders. Consumers are discovering and enjoying a wider variety of new and traditional beer styles—some brewed close to home and others brewed on the other side of the world. Traditional beer styles are not only celebrated in their native lands, but are now being shared, brewed and enjoyed worldwide. The number of

new styles and style interpretations appears to be limited only
by the creativity of the world's brewing community.

As consumer choices expand, the availability of information
becomes more important. With this in mind the Boulder,
Colorado–based Association of Brewers—an independent non-
profit company (which includes the American Homebrewers
Association, the Institute for Brewing Studies, Brewers
Publications, the International Beer Executives Symposium
and the Great American Beer Festival) with almost twenty
years experience in organizing successful beer competitions
and events, setting a standard for international competitions—developed and staged the first World Beer Cup International Competition in June 1996 in Vail, Colorado. The
competition will be held every two years. An experienced
professional panel of beer judges is chosen and charged
with honoring the top three beers in each of the style categories (in 1996 there were sixty-one of them) with gold,
silver and bronze awards.

The mission of the World Beer Cup International Competition is to educate beer drinkers around the world about
the stylistic traditions, beer cultures, quality and diversity
displayed by the world's vast array of beers. The WBC is
dedicated to recognizing the traditions and innovations
shared by the entire brewing industry as well as those customs unique to specific countries and regions.

By promoting today's international brewing activities, the
World Beer Cup aims to enhance consumer understanding
that there are no borders limiting the craft, art and science
of brewing. The WBC also actively promotes the responsible
consumption of beer as an alcohol-containing beverage, and
urges beer producers and distributors to encourage beer
enthusiasts to savor the flavor of beer responsibly.

WORLD BEER CUP—THE EVENT

More than 3,000 breweries worldwide were sent invitations to participate in the inaugural 1996 World Beer Cup International Competition. More than 250 breweries responded from every continent, entering a total of more than 600 different beers. Breweries and importers were responsible for entering their beers in the categories they considered most appropriate; competition organizers were not involved in these decisions. In most cases the beers were shipped from the breweries directly to the World Beer Cup's refrigerated warehouse site where they were sorted, coded and prepared for a blind judging.

Twenty-seven judges from breweries and companies around the world convened in Vail on June 13 and 14, 1996. Judges were selected from an Association of Brewers (also producers of Brewing Matters, Inc.'s The Great American Beer Festival) source list of internationally recognized brewers, consultants and writers. These individuals are selected on the basis of: (1) industry and peer recognition, (2) knowledge of beer styles and the brewing process, (3) flavor perception skills, (4) prior judging or beer evaluation experience, and (5) judging demeanor.

The judging took place in nine different sessions over the course of two days. Whenever possible, judges were assigned by World Beer Cup managers to judge beers in their specific area of expertise, though they were never allowed to judge a category in which they had entered beers or had an interest or concern. (For example, a consultant could not judge the beer of a client.)

Judges were divided into groups of three or more to evaluate numbered samples of beer. All beer was presented anonymously. Breweries could enter up to three of their products. Twenty-one competition captains and stewards prepared and presented the beers to the sequestered judges.

Typically no more than fifteen beers were presented in

any one round of judging. As an example, if a category had twenty-two entries, there would be two simultaneously judged rounds of eleven beers. Each set of three judges would choose the top three beers and nominate them to move forward to a second and final round, where usually all six judges evaluated the renumbered beers and chose Gold, Silver and Bronze winners.

Occasionally judges chose not to give an award in a category. The guidelines and World Beer Cup Awards philosophy used for determining awards are:

The World Beer Cup awards beers for excellence in sixty-one categories and does not automatically recognize the top three finishers in a particular category. When judges decide a category contains three excellent examples of the style, they present Gold, Silver and Bronze awards for the first-, second- and third-place beers, respectively. If judges believe that no beer entered in the category meets the quality and style-accuracy criteria, they may elect not to distribute any awards. Judges may recognize a beer as a Silver- or a Bronze-winning beer, yet not award a Gold.

The following World Beer Cup award criteria are the minimum standards of quality required for any beer considered for an award.

Gold
A world-class beer that accurately exemplifies the specified style, displaying the proper balance of taste, aroma and appearance.

Silver
An excellent beer that may vary slightly from style parameters while maintaining close adherence to the style and displaying excellent taste, aroma and appearance.

Bronze
A fine example of the style that may vary slightly from style parameters and/or have minor deviations in taste, aroma or appearance.

Judges may choose not to designate any awards in a category if all beers entered in that category are significantly out of style or have major defects, or if fewer than three entries were received.

The Winners of
The Association of Brewers
1996 World Beer Cup
International Competition

ALES

British Origin

Category 1:
Classic English-Style Pale Ale—12 Entries
<u>Gold</u>: No Gold Awarded
<u>Silver</u>: Pullman Pale Ale, Riverside Brewing Co., Riverside, California, USA
<u>Bronze</u>: Doggie Style Ale, Broadway Brewing LLC, Denver, Colorado, USA

Category 2:
India Pale Ale—26 Entries
Gold: Wet Mountain India Pale Ale, Il Vicino, Salida, Colorado, USA
Silver: Blind Pig India Pale Ale, Blind Pig Brewing Co., Temecula, California, USA
Bronze: Avery India Pale Ale, Avery Brewing Co., Boulder, Colorado, USA

Category 3:
English-Style Ordinary Bitter—2 Entries
Gold: Ruddles Best Bitter, Ruddles Brewery, Rutland, United Kingdom
Silver: Boddington's Bitter, Boddington's Brewery, Manchester, England (American Importer: LaBatt USA/InterBrew)
Bronze: No Bronze Awarded

Category 4:
English-Style (Special) Best Bitter—3 Entries
Gold: Delaney's Irish Ale, South China Brewing Co. Ltd, Aberdeen, Hong Kong
Silver: Galena ESB, Sun Valley Brewing Co., Hailey, Idaho, USA
Bronze: No Bronze Awarded

Category 5:
English-Style (Extra Special) Strong Bitter—19 Entries
Gold: Stoddard's ESB, Stoddard's Brewhouse and Eatery, Sunnyvale, California, USA
Silver: Extra Special Bitter Ale, Bison Brewing Co., Berkeley, California, USA
Bronze: Rikenjaks ESB, Rikenjaks Brewing Co., Inc. Lafayette, Louisiana, USA

Category 6:
Scottish-Style Ale—7 Entries
Gold: Scotch Ale, Samuel Adams Brewhouse, Philadelphia, Pennsylvania, USA

Silver: Laughing Lab Scottish Ale, Bristol Brewing, Colorado Springs, Colorado, USA

Bronze: Right Field Red Ale, Sandlot Brewery at Coors Field, Denver, Colorado, USA

Category 7:
English-Style Mild Ale—2 Entries
Gold: Seabright Session Ale, Seabright Brewery, Santa Cruz, California, USA
Silver: No Silver Awarded
Bronze: No Bronze Awarded

Category 8:
English-Style Brown Ale—15 Entries
Gold: Redwood Coast Brown Ale, Redwood Coast Brewing Co., Alameda, California, USA
Silver: Red Brick Ale, Atlanta Brewing Co., Atlanta, Georgia, USA
Bronze: Granville Island Pale Ale, Granville Island Brewing, Vancouver, British Columbia, Canada

Category 9:
Strong Ale—17 Entries
Gold: Star Brew 1000 Wheat Wine, Marin Brewing Co., Larkspur, California, USA
Silver: No Silver Awarded
Bronze: No Bronze Awarded

Category 10:
Barley Wine–Style Ale—9 Entries
Gold: Ozone Ale, Hubcap Brewery/Brewing Co., Vail, Colorado, USA
Silver: Big 12 Barley Wine, Little Apple Brewing Co., Manhattan, Kansas, USA
Bronze: Old Blue Granite Barley Wine, Blind Pig Brewing Co., Temecula, California, USA

Category 11:
Robust Porter—18 Entries
<u>Gold</u>: Saint Brigid's Porter, Great Divide Brewing Co., Denver, Colorado, USA
<u>Silver</u>: BridgePort Porter, BridgePort Brewing Co., Portland, Oregon, USA
<u>Bronze</u>: Edmund Fitzgerald Porter, Great Lakes Brewing Co., Cleveland, Ohio, USA

Category 12:
Brown Porter—7 Entries
<u>Gold</u>: St. Charles Porter, Blackstone Restaurant & Brewery, Nashville, Tennessee, USA
<u>Silver</u>: Haystack Black, Portland Brewing Co., Portland, Oregon, USA
<u>Bronze</u>: Stoddard's Porter, Stoddard's Brewhouse and Eatery, Sunnyvale, California, USA

Category 13:
Sweet Stout—3 Entries
<u>Gold</u>: Mackeson XXX Stout, Whitbread Beer Co., Bedfordshire, England (American Importer: Schoenling Brewing Co./USA)
<u>Silver</u>: Stillwater Cream Stout, Colorado Brewing Co., Thornton, Colorado, USA
<u>Bronze</u>: No Bronze Awarded

Category 14:
Oatmeal Stout—9 Entries
<u>Gold</u>: Zoser Oatmeal Stout, Oasis Brewery, Boulder, Colorado, USA
<u>Silver</u>: No Silver Awarded
<u>Bronze</u>: Gray's Classic Oatmeal Stout, Gray Brewing, Janesville, Wisconsin, USA

Category 15:
Imperial Stout—8 Entries
<u>Gold</u>: Old Rasputin Russian Imperial Stout, North Coast Brewing Co. Inc., Fort Bragg, California, USA

Silver: Imperial Stout, Wiibroes Brewery, Helsingor, Denmark

Bronze: Centennial Russian Imperial Stout, Trinity Brewhouse, Providence, Rhode Island, USA

Irish Origin

Category 16:

Classic Irish-Style Dry Stout—8 Entries

Gold: Founders Stout, Mishawaka Brewing Co., Mishawaka, Indiana, USA

Silver: Neptune Black Sea Stout, Neptune Brewery, New York, New York, USA

Bronze: Seminole Stout, Buckhead Brewery and Grill, Tallahassee, Florida, USA

Category 17:

Foreign-Style Stout—9 Entries

Gold: San Quentin's Breakout Stout, Marin Brewing Co, Larkspur, California, USA

Silver: Cascade Special Stout, Cascade Brewery, Hobart, Tasmania, Australia (American Importer: Carlton Special Beverage Co.)

Bronze: Echigo Stout, Uehara Shuzou, Echigo Beer, Nishikanbara-Gun, Japan

North American Origin

Category 18:

American-Style Pale Ale—26 Entries

Gold: Snake River Pale Ale, Snake River Brewing Co. Inc., Jackson Hole, Wyoming, USA

Silver: South Platte Pale Ale, Columbine Mill Brewery, Littleton, Colorado, USA

Bronze: Red Rocket Pale Ale, Bristol Brewing, Colorado Springs, Colorado, USA

Category 19:
American-Style Amber Ale—33 Entries
<u>Gold</u>: Capstone ESB, Oasis Brewery, Boulder, Colorado, USA
<u>Silver</u>: Devils Head Red, Columbine Mill Brewery, Littleton, Colorado, USA
<u>Bronze</u>: MacTarnahan's Ale, Portland Brewing Co., Portland, Oregon, USA

Category 20:
Golden Ale/Canadian-Style Ale—11 Entries
<u>Gold</u>: Griffon Extra Pale Ale, McAuslan Brewing Inc., Montréal, Québec, Canada
<u>Silver</u>: Molson Golden, Molson Breweries-MCI, Etobicoke, Ontario, Canada
<u>Bronze</u>: Independence Gold, Independence Brewing Co., Philadelphia, Pennsylvania, USA

Category 21:
American-Style Brown Ale—16 Entries
<u>Gold</u>: Slow Down Brown Ale, Il Vicino Inc., Albuquerque, New Mexico, USA
<u>Silver</u>: Saint Arnold Brown Ale, Saint Arnold Brewing Co., Houston, Texas, USA
<u>Bronze</u>: Nightwatch Dark Ale, Maritime Pacific Brewing Co., Seattle, Washington, USA

German Origin

Category 22:
German-Style Kölsch/Köln-Style Kölsch—4 Entries
<u>Gold</u>: Stoddard's Kölsch, Stoddard's Brewhouse and Eatery, Sunnyvale, California, USA
<u>Silver</u>: No Silver Awarded
<u>Bronze</u>: No Bronze Awarded

Category 23:
German-Style Brown Ale/Düsseldorf-Style Altbier—8 Entries
Gold: No Gold Awarded
Silver: Alle Tage Alt, McNeill's Brewery, Brattleboro, Vermont, USA
Bronze: Flagship Red Ale, Maritime Pacific Brewing Co., Seattle, Washington, USA

Category 24:
Berliner-Style Weisse (Wheat)—1 Entry
No Medals Awarded

Category 25:
South German–Style Weizen/Weissbier—22 Entries
Gold: Sundance Hefe-Weizen, Palmer Lake Brewing Co., Palmer Lake, Colorado, USA
Silver: Edelweiss Hefetrüb, Österreichische Bräu-Aktiengesellschaft, Linz, Austria (American Importer: Pa's Bier Inc./USA)
Bronze: Tabernash Weiss, Tabernash Brewing Co., Denver, Colorado, USA

Category 26:
South German–Style Dunkel Weizen/Dunkel Weissbier —4 Entries
Gold: Edelweiss Dunkel, Österreichische Bräu-Aktiengesellschaft, Linz, Austria (American Importer: Pa's Bier Inc./USA)
Silver: Paulaner Dunkel Weizen, Paulaner Brewery, Munich, Germany (American Importer: Paulaner North America/USA)
Bronze: No Bronze Awarded

Category 27:
South German–Style Weizenbock/Weissbock—2 Entries
Gold: Aventinus, Private Weissbierbrauerei G. Schneider & Son K.G., Munich, Germany (American Importer: G. Schneider & Son/ USA)
Silver: No Silver Awarded
Bronze: Wild Pitch Weizen Bock, Sandlot Brewery at Coors Field, Denver, Colorado, USA

Belgian and French Origin

Category 28:
Belgian-Style Flanders/Oud Bruin Ale—2 Entries
Gold: Liefmans Goudenband, Brouwerij Liefmans, Oudenaarde, Belgium (American Importer: Brewery Riva NV/USA)
Silver: No Silver Awarded
Bronze: Solstice Mystery Ale, Palmer Lake Brewing Co., Palmer Lake, Colorado, USA

Category 29:
Belgian-Style Abbey Ale—9 Entries
Gold: No Gold Awarded
Silver: Abbey Belgian Style Ale, New Belgium Brewing Co., Ft. Collins, Colorado, USA
Bronze: St. Bernardus Tripel, Brouerij St. Bernardus, Belgium (American Importer: Merchant du Vin Corp./USA)

Category 30:
Belgian-Style Pale Ale—2 Entries
Gold: No Gold Awarded
Silver: Fat Tire Amber Ale, New Belgium Brewing Co., Ft. Collins, Colorado, USA
Bronze: Orval Trappist Ale, Orval Trappist Monastery, Florenville, Belgium (American Importer: Merchant du Vin Corp./USA)

Category 31:
Belgian-Style Strong Ale—3 Entries
Gold: Pauwel Kwak, Brewery Bosteels, Buggenhout, Belgium
Silver: La Chouffe, Brasserie D'Achouffe, Achouffe, Belgium (American Importer: Dafoe Int'l/USA)
Bronze: No Bronze Awarded

Category 32:
Belgian-Style White (or Wit)/Belgian-Style Wheat—4 Entries
Gold: Hoegaarden White Beer, Brouwerij de Kluis, Hoegaarden, Belgium (American Importer: LaBatt USA/InterBrew)

Silver: Wit, Spring Street Brewing Co., New York, New York, USA
Bronze: Celis White, Celis Brewery Inc., Austin, Texas, USA

Category 33:
Belgian-Style Lambic—8 Entries
Gold: Lindeman's Cuvée René, Lindeman's Farm Brew, Blezenbeek, Belgium (American Importer: Merchant du Vin Corp./USA)
Silver: No Silver Awarded
Bronze: Belle-Vue Kriek, Brasserie Belle-Vue, Brussels, Belgium (American Importer: LaBatt USA)

Category 34:
French-Style Bière de Garde—1 Entry
Gold: Grain d'Orge, Brasserie Jeanne d'Arc, Ronchin-Lille, France
Silver: No Silver Awarded
Bronze: No Bronze Awarded

LAGER BEERS

European-Germanic Origin

Category 35:
German-Style Pilsener—11 Entries
Gold: Redwood Coast Alpine Gold Pilsner, Redwood Coast Brewing Co., Alameda, California, USA
Silver: Bohemia Beer, Cerveceria Cuauhtèmoc, Monterrey, Mexico (American Importer: LaBatt, USA)
Bronze: McClintic Pilsner, B.T. McClintic Brewing Co., Janesville, Wisconsin, USA

Category 36:
Bohemian-Style Pilsener—10 Entries
Gold: Ruffian Pilsner, Mountain Valley Brew Pub, Suffern, New York, USA
Silver: Brooklyn Lager, Brooklyn Brewery, Brooklyn, New York, USA

<u>Bronze</u>: *Bohemian Pilsner, Bohemian Brewery, Torrance, California,* USA

Category 37:
European-Style Pilsener—12 Entries
<u>Gold</u>: *Dos Equis Special Lager, Cerveceria Cuauhtèmoc, Monterrey, Mexico (American Importer: LaBatt USA)*
<u>Silver</u>: *Michael Shea's Blonde Lager, HighFalls Brewing Co., Rochester, New York, USA*
<u>Bronze</u>: *Efes Pilsen Erciyas, Biracilik Ve Malt San, A.S., Istanbul, Turkey*

Category 38:
European-Style Low-Alcohol Lager/German-Style Leicht— 2 Entries
<u>Gold</u>: *No Gold Awarded*
<u>Silver</u>: *Egger Leicht, Privatbrauerei Fritz Egger G.m.b.H, St. Pölten, Austria*
<u>Bronze</u>: *No Bronze Awarded*

Category 39:
Münchner-Style Helles—13 Entries
<u>Gold</u>: *Bow Valley Premium Lager, Bow Valley Brewing Co., Canmore, Alberta, Canada*
<u>Silver</u>: *Granville Island Lager, Granville Island Brewing, Vancouver, British Columbia, Canada*
<u>Bronze</u>: *Scrimshaw Beer, North Coast Brewing Co. Inc., Fort Bragg, California, USA*

Category 40:
Dortmunder/European-Style Export—8 Entries
<u>Gold</u>: *Stoudt's Export Gold, Stoudt's Brewing Co., Adamstown, Pennsylvania, USA*
<u>Silver</u>: *Berghoff Original Lager Beer, Joseph Huber Brewing Co., Monroe, Wisconsin, USA*
<u>Bronze</u>: *BCC, LECH Browary Wielkopolski SA, Poznan, Poland*

Category 41:

Vienna-Style Lager—7 Entries

Gold: Leinenkugel's Red Lager, Jacob Leinenkugel Brewing Co., Chippewa Falls, Wisconsin, USA

Silver: Dos Equis Amber Lager, Cerveceria Cuauhtèmoc, Monterrey, Mexico (American Importer: LaBatt USA)

Bronze: Blue Ridge Amber Lager, Frederick Brewing Co., Frederick, Maryland, USA

Category 42:

German-Style Märzen/Oktoberfest—6 Entries

Gold: No Gold Awarded

Silver: Paulaner Oktoberfest, Paulaner Brewery, Munich, Germany (American Importer: Paulaner North America/USA)

Bronze: Brasal Special Amber Lager, Brasserie Brasal Brewery, LaSalle, Québec, Canada

Category 43:

European-Style Dark Lager—11 Entries

Gold: Tabernash Munich Dark Lager, Tabernash Brewing Co., Denver, Colorado, USA

Silver: Blue Hen Black & Tan, Blue Hen Beer Co. Ltd., Newark, Delaware, USA

Bronze: Stovepipe Porter, Otter Creek Brewing Inc., Middlebury, Vermont, USA

Category 44:

German-Style Bock Beer—10 Entries

Gold: Ruffian Mai-Bock, Mountain Valley Brew Pub, Suffern, New York, USA

Silver: Hübsch Mai Bock, Sudwerk Privatbrauerei Hübsch, Davis, California, USA

Bronze: Augsburger Dopplebock, Augsburger Brewing Co., Detroit, Michigan, USA

Category 45:
German-Style Strong Bock Beer—9 Entries
<u>Gold</u>: Derailer Doppelbock, Tabernash Brewing Co., Denver, Colorado, USA
<u>Silver</u>: Scapegoat Doppelbock, Libertyville Brewing Co., Libertyville, Illinois, USA
<u>Bronze</u>: Bayrisch G. Frorns Eisbock, Kulmbacher Reichelbraeu, Kulmbach, Germany (American Importer: B. United/USA)

North American Origin

Category 46:
American-Style Lager—12 Entries
<u>Gold</u>: OB Lager, Oriental Brewery Co. Ltd., Seoul, Korea
<u>Silver</u>: Nex, Oriental Brewery Co. Ltd., Seoul, Korea
<u>Bronze</u>: Labatt Blue, Labatt Breweries of Canada, London, Ontario, Canada

Category 47:
American-Style Light Lager—5 Entries
<u>Gold</u>: Miller Lite, Miller Brewing Co., Milwaukee, Wisconsin, USA
<u>Silver</u>: Pabst Genuine Draft Light, Pabst Brewing Co., Milwuakee, Wisconsin, USA
<u>Bronze</u>: Medalla Light Beer, Cerveceria India Ale Inc., Mayaguez, Puerto Rico

Category 48:
American-Style Premium Lager—14 Entries
<u>Gold</u>: Brick Red Baron, Brick Brewing Co. S. Waterloo, Ontario, Canada
<u>Silver</u>: Signature, Stroh Brewery Co., Detroit, Michigan, USA
<u>Bronze</u>: Budweiser, Anheuser-Busch Inc., St. Louis, Missouri, USA

Category 49:
Dry Lager—1 Entry
<u>Gold</u>: No Gold Awarded
<u>Silver</u>: No Silver Awarded

Bronze: *Cerveja Antarctica, Cia. Antarctica Paulista, Jaguariúna, São Paulo, Brazil*

Category 50:
American-Style Ice Lager—4 Entries
Gold: *Icehouse, Plank Road Brewery, Milwaukee, Wisconsin, USA*
Silver: *Molson Ice, Molson Breweries-MCI, Etobicoke, Ontario, Canada*
Bronze: *Schlitz Ice, Jos. Schlitz Brewing Co., Detroit, Michigan, USA*

Category 51:
American-Style Malt Liquor—3 Entries
Gold: *Olde English 800 Malt Liquor, Pabst Brewing Co., Milwaukee, Wisconsin, USA*
Silver: *Schlitz Malt Liquor, Jos. Schlitz Brewing Co., Detroit, Michigan, USA*
Bronze: *Country Club Malt Liquor, Pearl Brewing Co., San Antonio, Texas, USA*

Category 52:
American-Style Amber Lager—14 Entries
Gold: *Point Amber Lager, Stevens Point Brewery, Stevens Point, Wisconsin, USA*
Silver: *JJ Wainwright Evil Eye Amber Lager, Pittsburgh Brewing Co., Pittsburgh, Pennsylvania, USA*
Bronze: *Red Wolf, Anheuser-Busch Inc., St. Louis, Missouri, USA*

Category 53:
American-Style Dark Lager—1 Entry
No Medals Awarded

Other Origin

Category 54:
Tropical-Style Light Lager—3 Entries
Gold: *Cascade Pale Ale, Cascade Brewery, Hobart, Tasmania, Australia (American Importer: Carlton Special Beverage Co./Australia)*

Silver: No Silver Awarded
Bronze: No Bronze Awarded

HYBRID/MIXED STYLES

Category 55:
American-Style Lager/Ale or Cream Ale—5 Entries
Gold: *California Blonde Ale, Coast Range Brewing Co., Gilroy, California, USA*
Silver: No Silver Awarded
Bronze: *Point Pale Ale, Stevens Point Brewery, Stevens Point, Wisconsin, USA*

Category 56:
American-Style Wheat Ale or Lager—16 Entries
Gold: *Thomas Kemper Hefeweizen, Thomas Kemper Brewing Co., Seattle, Washington, USA*
Silver: *Weiss Guy Wheat, Alcatraz Brewing Co., Indianapolis, Indiana, USA*
Bronze: *Red Ass Honey Wheat, Red Ass Brewing Co., Ft. Collins, Colorado, USA*

Category 57:
Fruit Beers—20 Entries
Gold: *Liefmans Frambozen, Brouwerij Liefmans, Oudenaarde, Belgium (American Importer: Brewery Riva NV/USA)*
Silver: *Belgian Red Wisconsin Cherry, New Glarus Brewing Co., New Glarus, Wisconsin, USA*
Bronze: *Liefmans Kriek, Brouwerij Liefmans, Oudenaarde, Belgium (American Importer: Brewery Riva NV/USA)*

Category 58:
Herb and Spice Beers—9 Entries
Gold: *Coriander Rye Ale, Bison Brewing Co., Berkeley, California, USA*
Silver: *RSB Spiced Scotch Ale, Routh Street Brewery, Dallas, Texas, USA*

Bronze: Wit Amber, Spring Street Brewing Co., New York, New
      York, USA

Category 59:
Specialty Beers—13 Entries
Gold: B&H Breakfast Toasted Ale, Barley & Hopps, San Mateo,
      California, USA
Silver: Schierlinger Roggen, Brauerei Thurn & Taxis, Schierling, Ger-
      many (American Importer: B. United/USA)
Bronze: Brewery Hill Honey Amber (Ale), The Lion Brewery, Wilkes-
      Barre, Pennsylvania, USA

Category 60:
Smoke-Flavored Beers (Ales or Lagers)—7 Entries
Gold: Aecht Schlenkerla Rauchbier, Brauerei Heller-Trum, Bamberg,
      Germany (American Importer: B. United/USA)
Silver: Alaskan Seasonal Smoked Porter, Alaskan Brewing & Bottling
      Co., Juneau, Alaska, USA
Bronze: No Bronze Awarded

Category 61:
Nonalcoholic Malt Beverages—5 Entries
Gold: Radegast Birell, Radegast Brewery J.S.C., Nošovice, Czech Republic
Silver: No Silver Awarded
Bronze: No Bronze Awarded

Tasting Gold

The long, involved process of formulating these recipes, beer-tasting notes and brewery notes involved a lot of historical and technical data collection, tasting and laboratory analysis. A summary of my journey will help lend perspective to how my perceptions and conclusions were reached.

TASTING THE BEERS

It took two months to taste all of the beers. The Gold Cup winners were evaluated first, followed by the random sampling of the Silver and Bronze winners. The beers were always kept under refrigeration at 36–40 degrees F (2–4.5 C) while in my care.

I took great care to taste each beer under identical situations. I tasted every beer from the same style of glass, a

500-ml. (about one pint) Belgian-style beer "snifter." A typical serving was three to four ounces, which allowed space
for aroma to develop within the glass and also raised the
temperature of the beer from 40 degrees F (4.5 C) to 45 to
50 degrees F (7.5–10 C), an optimal tasting temperature for
most beers. I tasted certain lighter lagers at colder temperatures and some specialty ales at warmer temperatures.
These differences are often noted in my evaluations.

   Never did I evaluate beers within two hours of having had
a meal. Ideally I sought to find late morning and predinner
times to exercise my tastings. I must confess, I am human
and not a machine. The words I chose to describe the beers
often reflected my mood at the time: *mechanical, sensual, joyous, happy, bored, poetic, creative, uninspired* and others. But the
attitude with which we all enjoy our beer is like life with all
its moods. And our human perceptions can never be removed from the biases, small and large, that our moods
may create. There will be agreement, concurrence and dis-
agreement with my assessments of these beers, based on
your own impressions, perspectives and perceptions of the
information I present here. After all, we're all beer experts,
right? What I have attempted to do is to realize and under-
stand the influences that affect my judgment and to establish
a routine for evaluating these beers to most consistently
present my own perceptions. How much did I drink? Usually never more than two ounces of any beer, unless it
was one I really enjoyed, in which case I would recap the
bottle and enjoy the rest later that evening, often with
friends (other than my friend the tree).

FURTHER EVALUATION

I had arranged for two of each of the winning beers to be
reserved from the competition. I wish I could tell you that
I had a great deal of pleasure enjoying the second bottle

of each beer at my leisure, but that was not the case. For an added perspective I sent all the gold winners to a laboratory for analysis using a sophisticated piece of equipment called SCABA for determining original extract, alcohol, apparent extract and color. A separate chemical analysis was made to determine IBUs (International Bittering Units). Laboratory analysis is not some special privilege I had access to; it is accessible to anyone. You can send your own beer or any beer off the shelf of your local store.

Finally, letters and forms were sent to each of the winning breweries requesting historical and technical information regarding the original extract, alcohol, apparent extract, color, bittering units, type and amount of ingredients used. Realizing that some of this information may have been considered proprietary by the brewery, I anticipated the divulgence of only a limited amount of technical information. I know that "a recipe does not a beer make," but I felt fortunate to obtain nearly a 100 percent response from the Gold Cup winners of the competition. Many of the responses included great detail, while others provided very basic information that proved invaluable along with the laboratory analyses, in fine-tuning my own assessments.

In most cases laboratory analysis provided supporting evidence of my own assessments and those of the breweries. Even though lab analyses are accurate and reliable, I did not always take them to be exact representations of the beers I was tasting. My goal with this book was to assess these beers and formulate homebrew recipes primarily based on how the beers tasted. Sometimes the analysis deviated from the brewery's assessment and my own perceptions. Compromises were made between the three sets of observations with the question in mind, "What is going to work for a homebrewer making beer five gallons at a time?"

One very important piece of information I learned regards the misrepresentation inherent in some of the standardized laboratory analysis for color determination. The intensity, darkness and lightness of color one visually perceives does not consistently conform to the SRM (Standard Reference Method) of analysis. For example, beers that were clearly in the 10 to 17 degrees Lovibond visual comparison scale (often equated to the SRM color scale) were analyzed to be significantly darker. The SRM of color determination measures the intensity of certain wavelengths of light. It does not take into consideration the influence of some red and orange hues and how color is actually perceived by the human eye. The analysis is accurate and the numbers are accurate, but they do not always match the degree of light intensity we actually see with our eyes. There were clearly many anomalies, and in these instances my own visual assessment became the data represented in this book. Interestingly, it often turned out that my own visual assessment and color assignments more closely coincided with grain and extract formulation and predicted color of the beer. Also, the brewery's own recipe formulations more often coincided with predicted color based on proportions of the amounts of colored specialty malts than they did with laboratory analysis. The SRM of color analysis is only a valid method for beers in the 2 to 5 or 6 range. I believe the use of specialty malts that contribute red and orange hues confuses the method and consequently provides data that are inconsistent with visual assessment.

A WORD ABOUT TARGETS

You will note a set of targets for each recipe in terms of original specific gravity, final gravity, alcohol, color and bittering units. These data are not an exact representation of any one means of evaluation, nor are they an average. Brewery data, lab analysis and my own sensory assessments were compared with each other. What my palate perceived, my eyes saw, my mouth felt and my nose smelled were all important considerations in representing the targets for homebrewing 5-gallon (19-l.) batches of these beers.

At times I chose to omit certain ingredients that the breweries indicated they had used. Sometimes I added an ingredient the brewery did not use. These changes were made based on my assessment of color, aroma, taste or mouth feel in the interest of achieving the character I perceived in the beer I tasted—the beer that won the 1996 World Beer Cup.

Some breweries brew a quality product that is always consistent in character. Other breweries brew a quality product that varies slightly from batch to batch. All beers change with time and are influenced by handling and the conditions under which they are stored. The recipes presented here closely match the actual beers that were entered, judged and further evaluated for the World Beer Cup and *Home Brewer's Gold*.

WORLD
BEER CUP

Homebrewing and Your Expectations: Things You Need to Know Before Using These Recipes

A recipe does not a beer make.

Brewing 1,000 barrels (1,200 hl.) at a time is extraordinarily different from homebrewing 5 gallons (19 l.). There's even a significant difference between 5 gallons (19 l.) and 200 gallons (760 l.). Those differences are a result of ingredient and material handling, equipment efficiencies, fermentation behavior, packaging stress and ability to control the process. These differences are discussed in detail in my *Home Brewer's Companion*. Briefly presented here are discussions of some of the most important considerations to keep in mind when using the recipe formulations in this book.

There are all manners of styles and setups for homebrew

equipment. Your equipment and your methods for using it are unique and will invariably influence the final character of your beer. That recognized—all of the recipes are based on a standard set of assumptions and methods. You will need to take these into consideration in order to formulate the recipe *you* use based on the efficiencies of your equipment and process.

1. Mash efficiency is assumed to be 75 percent. This means that of the potential maximum yield you might expect from your grains, the recipes assume you will only be able to attain 75 percent of that maximum. If, for example, you know that you can achieve greater efficiency, then you should use proportionally less grain or produce a higher volume of beer (add more water—*more beer*!) to achieve the same extract from the grains.

2. Boiling times of more than 60 minutes are approximations. Often more boiling time is needed to evaporate water for high-gravity beers. The energy output of your heat source will dictate your total boiling time.

3. Metric conversions are made from English units and are rounded off to the nearest significant decimal point. When there is a discrepancy between accurate and approximate conversion, regard the difference as insignificant in the big picture.

4. Conversion to EBC visual color units from SRM visual color units was achieved simply by multiplying SRM by 2. There is no consistent factor for converting SRM malt color units to EBC malt color units and vice versa. EBC malt color units are determined by a different method than SRM, and there is not a linear or regular relationship between the two.

5. Hop utilization is influenced by the form of hops, quality of the boil and the density of wort in which the hops are boiled. All these are taken into consideration for a wort that will yield 5 gallons. All recipes have been formulated with hop form specifically designated. Hop pellets can always be substituted for whole hops and vice versa, but a difference in utilization rates must be taken into consideration. If pellets are substituted for whole hops, use 15 percent less than indicated in the recipe. If whole hops are substituted for pellets, use 15 percent more than indicated in the recipe. For a full-wort one-hour (or more) boil of 1.040–46 (10–11.5) gravity beer, 30 percent utilization is assumed with hop pellets. Fifteen percent less or 25.5 percent utilization would be assumed if using whole hops. As the density of the boiled wort is increased, hop utilization decreases. This has been taken into account in formulating the recipes.

6. HBUs (Homebrew Bitterness Units, sometimes referred to as AAUs or Alpha Acid Units) is a quantitative measure of hops convenient for homebrew formulations. Homebrew Bitterness Units represents the number of ounces of hops multiplied by their percent alpha acid rating. One ounce of a 5 percent alpha acid hop is equal to 5 HBUs. One half ounce of a 10 percent alpha acid hop is also equal to 5 HBUs.

MBUs is a new unit introduced in this book. Metric Bitterness Units is equal to the number of grams of hops multiplied by the percent alpha acid rating. One ounce is equal to 28.35 grams. One HBU = 28 MBU. Ten grams of 5 percent alpha acid hops is equal to 50 MBU. Five HBU = $28.35 \times 5 = 141.75 = 142$ MBU.

7. Quality of water is of great importance in brewing beers to style or type. Information regarding this subject is widely available in brewing literature. Recommended readings on this subject are Ray Daniels' *Designing Great Beers* (Brewers Publications, 1996) and Gregory Noonan's *The New Brewing Lager Beers* (Brewers Publications, 1996). Mineral content and pH are critical when attempting to perfect a specific beer or beer style. Generally speaking, lighter lagers are best brewed with soft water with low mineral content. Nearly all commercially made beers are brewed with water that is not excessively alkaline nor high in pH. Exceptions can be made with beers using roasted grains. The quality of some classic British ales is sometimes attributable to water high in sulfates. When in doubt, begin with soft water having neutral pH.

8. Oftentimes breweries will use yeast that is commercially available to anyone. If this information was divulged by the brewery, then this is the yeast recommended. In many instances brewers use proprietary yeast, not commercially available. Under these circumstances yeasts are recommended based on my own experiences or assessment. There are several laboratories that culture, package and sell yeast to the homebrewer. Wyeast Laboratory yeasts are suggested in most recipes because of their widespread availability. Their use is not intended to be an endorsement of one brand over another. Other brands of yeast can be substituted interchangeably if availability or preference dictates.

9. For convenience and quality, most homebrewers bottle-condition their beers. Unless noted, most of the winning beers underwent some type of filtration and carbonation prior to packaging. The difference in the quality will be notable, but the overall character of your homebrewed version will be faithful to the commercial product in most cases. One can always exercise the option to filter and carbonate the beer prior to bottling to more closely finesse the detailed character of the winning beer.

10. Many of the winning beers are pasteurized in order to help stabilize the flavor for a longer period of time. This process has some effect on the overall character. Your unpasteurized homebrew will be different. As a homebrewer myself, I believe the character of unpasteurized beer, when the beer is cared for properly, is better.

11. Additional finesse. The process of whirlpooling and kraeusening outlined in more detail in many homebrew books, including *The Home Brewer's Companion*, can help you to more closely match the character of the winning beers. Whirlpooling removes protein trub from the boiled wort before it is chilled. Removing trub before fermentation will give the beer an added cleanness and greater flavor stability. Virtually all commercial brewers practice some form of trub removal, usually by whirlpooling (swirling) the hot wort in a tank and drawing off clear wort from the side, while the trub sediment migrates naturally to the center of the tank.

Rarely do commercial brewers use sugar to prime or carbonate their beers. Instead, they use kraeusening methods, which involve adding malt extract or new wort in measured amounts to finished beer. Refer to my book *The New Complete Joy of Home Brewing* for simple kraeusening procedures for homebrewers.

12. No winning recipe used malt extract as the basis for beer. Malt-extract and mash-extract recipes have been formulated for the beginning and intermediate homebrewer wishing to participate in the adventure of brewing world-class beers. The malt-extract versions will be damned close and certainly every bit as satisfying as an all-grain recipe, but in many cases will not as closely match the character of the commercial/winning beer. You will often find me brewing mash-extract versions due to time and effort limitations. I personally would not hesitate to brew an extract or mash-extract version; in fact, my preference is for mash-extract procedures.

13. There are hundreds of malt extracts available at your local homebrew supply store. Some can be used interchangeably with the dried malt extract called for in the recipes. English dried malt extract is chosen as the base for most malt-extract or mash-extract recipes because (1) it is my personal preference, (2) attenuation is excellent, (3) it can be conveniently measured, and the unused portion easily stored for future use, (4) it is some of the lightest malt extract available, and (5) the quality of dry malt is more stable with age.

Malt syrup can be substituted in many cases. Use 18 percent more malt extract syrup than dried extract. For example, if the recipe calls for 3 pounds dried extract, substitute 3½ pounds malt extract syrup.

14. Disclaimer: All of the recipes are my own formulations based on tasting notes, technical analysis, a limited amount of brewery information, and my own experiences as a homebrewer. In no way should any of these recipes and procedures be construed to represent the brewery's actual formulation and process.

15. Remember, you are a homebrewer. Have fun. Relax. Don't worry, and have a homebrew.

WORLD
BEER CUP

A Few Conclusions

The privilege of being able to carefully taste 146 winning beers was a special and unique experience. At first, though, I really didn't appreciate what a great opportunity I had before me. It seemed a disciplined task (yeah, right, tough job, but someone had to do it), an insurmountable pile of beer to be sorted, tasted, prodded and evaluated for this project. It wasn't until I had tasted about thirty or forty beers that I began to realize how different this tasting was from any other I had conducted before. Here were 146 beers judged the best by an extremely competent panel of professionals from all over the world. No judge or other person had ever tasted all of the winners before.

By beer number thirty or forty I began to realize there was a common thread with all of the winning beers, no matter what the style. Before me I had an answer to the question "What makes a beer a winner?"

Judges evaluate beers according to adherence to style as

well as the beer's overall cleanness. Much to the judges' credit, nearly all of the first-place beers were perfectly matched to the style descriptors—at least in accordance with sensory assessments. The most important realization for me was that every single beer was clean. Odd or exaggerated fermentation and packaging characters (diacetyl, acetaldehyde, DMS [dimethylsulfide], obtrusive esters, higher-solventlike alcohols, bacterial byproducts, oxidation, staling compounds, light-struck character, etc.) were totally absent in the Gold Cup beers, except where appropriate in certain styles. It was the malt and hops that were the primary indicators of character, with the light character of fermentation often contributing to the winning edge.

Silver and Bronze Cups were awarded to many beers that could have won the Gold. They had the minimum standards of a Gold Cup beer. But when minimum Gold standards were not perceived, often judges would drift toward choosing a second- or third-place winner on the merits of a beer's cleanness, rather than perfect adherence to style. It seems judges were more flexible and forgiving when a beer appeared a little out of style than they would be for a beer with fermentation or processing flaws.

It was fun for me to compare my sensory evaluation skills to brewery data and laboratory analysis. I learned a lot about my biases and have since attempted to readjust some of my sensory processing skills.

In most cases there was nothing really surprising in the brewery recipe formulations. Hops, malt and yeast do not a beer make. These words could not be truer. These winning beers and the beers you will brew will be a further testament to the skill of the brewer; your skill as a homebrewer. It's the passion, the love, the craft, the skill, the life the brewer gives to a written list of ingredients that really are the true measure of a winner. In these ways, the winners of the 1996 World Beer Cup represent the pride we all have as homebrewers.

The Beers, the Breweries, and the Recipes

For each category a style description is given, worded as it was presented to the entering breweries. These descriptions will evolve. One item of note is the contradiction of converting SRM color to EBC color in the recipes and the style descriptors. In the future a more accurate and straightforward factor of two will be used to convert SRM colors to EBC in the style descriptions. The contradiction arose when old methods of EBC color determination were confused with newer methods when defining each category. Therefore the EBC color rating in the style descriptions may be slightly inaccurate.

Brewery information and historical data for all the Gold winners of the 1996 World Beer Cup were provided by the breweries or their importers. The short descriptions are condensations of interesting anecdotes and significant historical milestones.

As previously stated, the taste descriptions are entirely my own and were made prior to reading any of the laboratory analyses or technical data provided by the breweries.

For the most part, detailed technical data were not requested of the Silver and Bronze winners. When available, the information is noted; otherwise, technical information is indicated as a reasonable guess based on personal evaluation.

Ales

British Origin

CATEGORY 1:
CLASSIC ENGLISH-STYLE PALE ALE

Classic English Pale Ales are golden to copper-colored and display English variety hop character. High hop bitterness, flavor and aroma are evident. This medium-bodied pale ale has low to medium maltiness. Low caramel is allowable. Fruity ester flavors and aromas are moderate to strong. Chill haze is allowable at cold temperatures. Diacetyl (butterscotch character) should be at very low levels or not perceived.

Original Gravity (°Plato):   1.044–1.056 (11–14 °Plato)

Apparent Extract–
Final Gravity (°Plato):   1.008–1.016 (2–4 °Plato)

Alcohol by weight (volume):   3.5–4.2% (4.5–5.5%)
Bitterness (IBU):   20–40
Color SRM (EBC):   4–11 (10–25 EBC)

Gold Cup Winner

None

Silver Cup Winner

Pullman Pale Ale
Riverside Brewing Co.
Riverside, California, USA

Medium amber color with orange hue. Pleasant alelike fruitiness combines with caramel and toasted malt and mild
earthy hop aroma. Flavor immediately impresses as well
balanced, pleasantly combining soft hop bitterness and mild
malt sweetness, with a slight crystal/caramel malt accent.
Hop flavor is also a significant part of the overall flavor
profile. An excellent rendition of this classic style. Medium-
bodied mouth feel with a clean finish that does not satiate
the palate. As an enticement for future indulgences, pleasant soft hop bitterness lingers in the aftertaste. Malts, hops
and fermented characters blend together, joining in an overall pleasant quenching sensation. Not too bitter but bitter
enough to justify it as a classic in the style. A simple beer
artfully made that says a lot.

Estimated profile based on tasting
Color: 12–14 SRM (24–28 EBC)
Bittering Units: 34–39

Bronze Cup Winner

Doggie Style Ale
Broadway Brewing LLC
Denver, Colorado, USA

Amber color with a reddish orange hue. The way the bubbles behave and slowly swirl around the glass and surface as the beer is poured suggests a full-bodied beer. Pleasant aroma of caramel and slight diacetyl are the primary aromas. Hops then emerge behind the malt character. Mouth feel is medium- to full-bodied, followed by something more than a dry finish. Malt character is primarily stated in the fullness of body, rather than flavor. An excellent clean bitterness impresses with its soft assertiveness. Flavor/aroma hops are probably used to bitter, flavor and lend aroma. Aftertaste lingers as a gently assertive bitterness, beckoning further indulgence. Promotes conversation about important things in life; a conversational ale, not outspoken in any singular way. Finishing hop aroma reminisces as floral-sweet. Clean with indefinite yet characteristic complex fruitiness. Brewery uses a combination of American Cascades and Northern Brewer.

Estimated profile based on tasting
Color: 12–14 SRM (24–28 EBC)
Bittering Units: 35–38

CATEGORY 2:
INDIA PALE ALE

India Pale Ales are characterized by intense hop bitterness with high alcohol content. A high hopping rate and the use of water with high mineral content result in a crisp, dry beer. This golden to deep copper-colored ale has a full, flowery hop aroma and may have a strong hop flavor (in addition to the hop bitterness). India Pale Ales possess medium maltiness and body. Fruity ester flavors and aromas are moderate to very strong. Chill haze is allowable at cold temperatures.

Original Gravity (°Plato):   1.050–1.070 (12.5–17.5 °Plato)
Apparent Extract–
Final Gravity (°Plato):   1.012–1.018 (3–4.5 °Plato)
Alcohol by weight (volume):   4–6% (5–7.5%)
Bitterness (IBU):   40–60
Color SRM (EBC):   8–14 (16–35 EBC)

Gold Cup Winner

Wet Mountain India Pale Ale
Il Vicino Wood Oven Pizza and Brewery
136 E. Second St.
Salida, Colorado, USA 81201
Brewmaster: Tom Hennessy;
Brewer: Jason Ackerman
Established 1994
Production: 550 bbl. (640 hl.)

The original Il Vicino in Albuquerque, New Mexico, was completed in 1992 and designed to capture the expressions of the pre–World War II period of city life by reinterpreting the northern Italian trattoria and the turn-of-the-century Paris bistro—all in an area of Albuquerque revitalized by renovation along Central Avenue, Route 66, the city's spine.

Il Vicino opened its second small Italian restaurant in the small town of Salida, Colorado, in 1994. The new Il Vicino, tucked away against a wedge of Rocky Mountains along the

Thomas J. Hennessy and Jason Ackerman, Brewers

Arkansas River, serves fine handmade Italian food in a neighborhood atmosphere.

Il Vicino ("the neighbor") is in part the concept of Tom Hennessy, who started the company in 1985 with partners Tom White, Greg Atkin, Rick Post and Ken Brock, in Albuquerque, New Mexico. One of Tom's hobbies took him into the realm of homebrewing, and soon the effects of good beer inspired the thought of installing a brewery in the Albuquerque restaurant. By the end of 1993 a very small 3-barrel brewery (later expanded to 7 barrels) had been installed, entirely hand-fashioned from used "Grundy" fermentation tanks from the U.K. The brewers fondly recall that "the brewery was an afterthought. We ripped out the dishwasher and built a ten-by-ten-foot brewhouse."

At the same time installation of another hand-fashioned brewery was under way at the sixty-seat restaurant in Salida. The system they fondly call "Frankenbrew" is a blend of creativity and necessity. Fermenters were turned into brew-kettles, mash vessels were made from dairy tanks, and fermenters were "stretched" with stainless steel welding magic to hold 3½ barrels of fermenting brew. They did it all for a total cost of less than $20,000 and a lot of sweat equity.

"Special attention is paid to the pH of the mash, wort and fermentation," says brewer-owner Hennessy. "We try to hit a target pH of the finished beer. Magic happens in our brewery when our pH is on target. But our secret in the Wet Mountain IPA is the blessing of the mash by certain particular tunes I play on the bagpipes."

Tom considers himself the head brewer for all the Il Vicino's recipes and brewery design, but gives due credit to his two brewers, Jason Ackerman in Salida and Brady McKeown in Albuquerque, for producing the two Gold Cup award-winning brews, Wet Mountain India Pale Ale (brewed in Salida) and Slow Down Brown Ale (brewed in Albuquerque). Beers that are a testimony to getting it right and making magic!

You'll find an assorted range of beers at both Il Vicinos, including Rob's Red, Joe Stout, Tad Raz Wheat, Old 66 Golden Ale, McKeown's Pale Ale, nonalcoholically brewed Root Beer and their award-winning Wet Mountain India Pale Ale and Slow Down Brown Ale.

CHARACTER DESCRIPTION OF GOLD CUP–WINNING WET MOUNTAIN INDIA PALE ALE

Orange-hued amber ale. Very pungent aroma presents a peppery, citrusy hop character with a suggestion of astringency. Ale fruitiness is masked by an intensive hop character. Malt sweetness underlying the hop aroma is not impressionable in overall assessment of aroma. Aroma misleads with the first taste. Flavor of hops is evident but hides behind a stylistically well-balanced, clean and very assertive hop bitterness. Bitterness is not entirely characteristic of a Centennial/Cascade type of hop. Information provided by the brewery confirms this impression. Bittering hops are a combination of American Northern Brewers and Cascade. Flavor and dry-hopping aroma hops are Centennial. The flavor character is reminiscent of black pepper, also suggested previously in aroma. The citrus character of Cascade and Centennial is well balanced for the style. Bitterness lies between a soft and harsh impact on the palate. Generally speaking, the malt character is subdued and in the background, but the contribution of medium body and rather well attenuated dry finish is adequate. Light touch of crystal malt comes through in the flavor. This beer is not biscuitlike in malt character nor excessively sweet. Its overall impression leaves you with a rather clean, light-colored pale malt sweetness.

- ### Recipe for 5 U.S. gallons (19 liters) Wet Mountain India Pale Ale

  Targets:
  Original Gravity: 1.060 (15)
  Final Gravity: 1.014 (3.5)
  Alcohol by volume: 6%
  Color: 14 SRM (28 EBC)
  Bittering Units: 60

ALL-GRAIN RECIPE AND PROCEDURE

9	lbs.	(4.1 kg.)	American 2-row pale malt
1	lb.	(0.45 kg.)	American Cara-Pils
½	lb.	(0.22 kg.)	American caramel—40 Lovibond
½	lb.	(0.22 kg.)	American caramel—80 Lovibond
11	**lbs.**	**(5 kg.)**	**Total grains**

2   HBU   (56 MBU) Cascade hops (pellets)—75 minutes (bittering)

7.7 HBU   (218 MBU) American Northern Brewers hops (pellets)—75 minutes (bittering)

11.5 HBU   (326 MBU) American Centennial hops (pellets)—20 minutes (flavor)

10   HBU   (284 MBU) American Centennial hops (pellets)—dry hop for 2 weeks (aroma)

¼   tsp.   Irish moss
¾   c.   corn sugar for priming in bottles. Use ⅓ cup
              corn sugar if priming a keg.
Wyeast 1187 Ringwood Ale yeast or Wyeast 1742 Swedish
Ale yeast

A step-infusion mash is employed to mash the grains.
Add 11 quarts (10.5 l.) of 138-degree F (58 C) water to the
crushed grain, stir, stabilize and hold the temperature at
128 degrees F (53 C) for 30 minutes. Add 5.5 quarts (5.2 l.)
of boiling water. Add heat to bring temperature up to 150
degrees F (65.5 C). Hold for about 60 minutes.

After conversion, raise temperature to 167 degrees F (75
C), lauter and sparge with 5 gallons (19 l.) of 170-degree F
(77 C) water. Collect about 6 gallons (23 l.) of runoff, add
bittering hops and bring to a full and vigorous boil.

The total boil time will be 75 minutes. When 20 minutes
remain, add flavor hops. When 10 minutes remain, add Irish
moss. After a total wort boil of 75 minutes (reducing the
wort volume to just over 5 gallons), turn off the heat, separate or strain out and sparge hops. Chill the wort to 65 to
70 degrees F (18–21 C) and direct into a sanitized fermenter.
Aerate the cooled wort well. Add an active yeast culture and
ferment for 4 to 6 days in the primary. Then transfer into a
secondary fermenter, chill to 55 to 60 degrees F (13–15.5
C) and add aroma-dry hops. Allow to age for two weeks
or more.

When secondary aging is complete, prime with sugar, bottle
or keg. Let condition at temperatures above 60 degrees F
(15.5 C) until clear and carbonated.

MALT-EXTRACT RECIPE AND PROCEDURE FOR WET MOUNTAIN
INDIA PALE ALE
5½   lbs.   (2.5 kg.)   British extralight dried malt extract
½   lb.   (0.22 kg.)   American caramel—40 Lovibond
½   lb.   (0.22 kg.)   American caramel—80 Lovibond
1   lb.   (0.45 kg.)   Total grains

2.5 HBU Cascade hops (pellets)—75 minutes (bittering)

13.5 HBU American Northern Brewers hops (pellets)—75 minutes (bittering)

11.5 HBU (326 MBU) American Centennial hops (pellets)—20 minutes (flavor)

10 HBU (284 MBU) American Centennial hops (pellets)—dry-hop for 2 weeks (aroma)

¼ tsp. Irish moss
¾ c. corn sugar for priming in bottles. Use ⅓ cup corn sugar if priming a keg.

Wyeast 1187 Ringwood Ale yeast or Wyeast 1742 Swedish Ale yeast

Steep crushed specialty grains in 1½ gallons (5.7 l.) water at 150 degrees F (65.5 C) for 30 minutes. Strain and sparge with enough 170-degree F (76.5 C) water to finish with a little over 2.5 gallons (9.5 l.) specialty grain liquor. Add the dried malt extract and bittering hops and bring to a full and vigorous boil.

The total boil time will be 75 minutes. When 20 minutes remain, add flavor hops. When 10 minutes remain, add Irish moss. After a total wort boil of 75 minutes, turn off the heat, separate or strain out and sparge hops, and direct the hot wort into a sanitized fermenter to which 2 gallons (7.6 l.) of cold water have been added. If necessary, add additional cold water to achieve a 5-gallon (19 l.) batch size. Chill the wort to 65 to 70 degrees F (18–21 C). Aerate the cooled wort well. Add an active yeast culture and ferment for 4 to 6 days in the primary. Then transfer into a secondary fermenter, chill to 55 to 60 degrees F (13–15.5 C) and add aroma-dry hops. Allow to age for two weeks or more.

When secondary aging is complete, prime with sugar, bottle or keg. Let condition at temperatures above 60 degrees F (15.5 C) until clear and carbonated.

Silver Cup Winner

Blind Pig India Pale Ale
Blind Pig Brewing Co.
Temecula, California, USA

Light amber ale color with orange hue. Aroma has plenty of hop character to stimulate and blast the senses. Strong, complex, floral, gentle hop aroma. American hops character showcased throughout, yet some aspects of the hop character hint at European-type nuances. A skillful treatment of pale malts lends a strong but not robust malt aroma. Then watch out and be prepared, for it's all hops in the flavor after only a brief visit with subtly sweet malt and light caramel characters. Bitterness seems at a level of 70 to 80, but the brewery claims 92! Though it's all American Chinook, Cascade, Columbus and Centennial bitterness, flavor and dry hopping, there is something here that resembles the character of an English-grown Kent hop. My palate fortunately won't bet on it; fooled by the craft. The overall bitterness is sharp, punctual and to the point. There is nothing gentle about the bitterness, suggesting a combination of medium and higher alpha hop. Aftertaste has a lingering sweetness with an overriding hop bitterness, never fading as it continues to charge with another wonderful assault on the palate. Even the belches and burps are a resurrection of hop aroma. Say amen, brothers and sisters. Nothing short of a hop lover's dream ale. Great balance of hop aroma, flavor and bitterness. Enough about hops; this beer is medium- to full-bodied with a clean (but hoppy) finish. Blind Pig India Pale Ale may have complex ale fruity esters, but they are overshadowed with hops, hops and more hops.

Estimated profile based on tasting
Original Gravity: 1.065–1.070 (16–17.5)
Final Gravity: 1.014–1.016 (3.5–4)
Alcohol by volume: 6–7%

Color: 7–9 SRM (14–18 EBC)
Bittering Units: 70–80 (92 indicated by the brewery)

Bronze Cup Winner

Avery India Pale Ale
Avery Brewing Co.
Boulder, Colorado, USA

Pale golden color with an orange hue. The aroma screams of American-grown hops and lots of them. A lot of fruity/citrus/grapefruitlike character with a bit of grassy hop essence. Caramel malt aroma tries to emerge, but really doesn't quite make it. No matter, for there's plenty of pale malt sweetness and a touch of caramel malt character in the flavor and full-bodied texture. Hops justify their existence emerging upon the palate, stampeding and thundering like a million buffalo on the plains of Colorado 200 years ago. Each drop of hoppy beer reaches the far corners of the palate. Then the refreshing fierceness strangely diminishes in the aftertaste. Now, don't get me wrong, the aftertaste is bitter, but the intensity does not linger or follow through as one might anticipate with the initial onslaught. Bitterness may be above the 70 or 80 range, but really now, who cares after that?

Brewery formulation uses Chinook for flavoring and bittering, while B.C. Canadian Kent Goldings are used as aroma and dry hops. A 10 percent combination of Munich and dark caramel malts is combined with pale malts for a 1.070 (17.5) original extract.

Estimated profile based on tasting
Original Gravity: 1.070 (17.5) indicated by the brewery
Final Gravity: 1.014–1.018 (3.5–4.5)
Alcohol by volume: 6–7%
Color: 9–11 SRM (18–22 EBC)
Bittering Units: 70–80

CATEGORY 3:
ENGLISH-STYLE ORDINARY BITTER

Ordinary Bitter is gold- to copper-colored with medium bitterness, light to medium body and low to medium residual malt sweetness. Diacetyl and fruity-ester properties should be minimized in this form of bitter.

Original Gravity (°Plato):	1.033–1.038 (8–9.5 °Plato)
Apparent Extract– Final Gravity (°Plato):	1.006–1.012 (1.5–3 °Plato)
Alcohol by weight (volume):	2.4–3.0% (3–3.7%)
Bitterness (IBU):	20–35
Color SRM (EBC):	8–12 (16–20 EBC)

Gold Cup Winner

Ruddles Best Bitter
Ruddles Brewery Limited
Langham, Oakham
Rutland LE15 7JD
England
Brewmaster: Alan Dunn
Established: 1858
Production: 342,000 bbl. (400,000 hl.)

The Ruddles Brewery, originally established by the Parry family as a small hometown brewery, was bought in 1911 by a member of the board, George Ruddle. From then until 1987 the Ruddles Brewery was a family-run, small, independent brewery. In the 1970s opportunities arose tempting the company to expand and make its products more widely available. In 1987 Grand Metropolitan, a large corporation with many brewery holdings, agreed to purchase the Rud-

Justine Fosh

Ruddles Brewery, c. 1900

dles Brewery. Ruddles continued to make premium and cask-conditioned real ale products. In 1989 with legislation pending that would affect the brewing industry, Grand Metropolitan sold Ruddles to Courage. Ruddles's distribution continued to grow, while it continued to produce real ale, along with other packaged products.

The Ruddles story continued with its purchase by the Dutch brewing company Grolsch in 1992. Two years later Bass Brewers was sold the right to market and sell Ruddles under a joint venture between Grolsch and Bass.

Now the brewery claims it operates as a separate, stand-alone regional brewing company. Since its origins as a tiny local "microbrewery" in 1858, the company has managed to maintain the qualities of the best of the British brewing tradition. Some may wonder whether all this business of mergers, buyouts and trades really mattered. Well, yes, it did; otherwise we wouldn't have the opportunity to enjoy such a world-class beer in so many convenient places.

What's on tap from Ruddles? Besides their award-winning Best Bitter, you can often also enjoy the company of Ruddles Strong County Ale.

CHARACTER DESCRIPTION OF GOLD CUP–WINNING RUDDLES BEST BITTER

An orange-hued amber color reminiscent of Oktoberfest beer. The rich, creamy head is spectacular due to the "widget" that activates a release of nitrogen gas within the can when it is opened, creating a dense and sensual foam. There is little doubt that real Kent Goldings hops create a predominant overture in the aroma. Malt aroma is absent, perhaps due to such a wonderful fullness of hop aromatics. Or perhaps the thick head of foam creates a barrier of sorts, preventing the emergence of malt character from the beer beneath. Yes, perhaps, but a pleasant barrier between beer and me. No fruitiness or fermentation character noted. No diacetyl evident. First flavor impression provokes thoughts of exceptional smoothness with a mild hop flavor and pleasant, gentle bite to the bitterness. Bitterness is relatively low but distinctive. Aftertaste is hop flavor and bitterness. Very little caramel or malt in flavor, though a foundation of crystal or caramel malt can be perceived, serving to support the hop bitterness and flavor. Goldings is on the mind, but if my hunch is correct, it isn't all Goldings. Overall, well-balanced, extraordinarily clean aftertaste and relatively mild palate for the category. Great drinkability and session beer.

- **Recipe for 5 U.S. gallons (19 liters) Ruddles Best Bitter**
  Targets:
  Original Gravity: 1.036 (9)
  Final Gravity: 1.008 (2)
  Alcohol by volume: 3.6%
  Color: 12 SRM (24 EBC)
  Bittering Units: 30

ALL-GRAIN RECIPE AND PROCEDURE

6¾ lbs.	(3.1 kg.)	English 2-row pale malt
¼ lb.	(114 g.)	English crystal malt—120 Lovibond
7 lbs.	**(3.2 kg.)**	**Total grains**

4   HBU   (113 MBU) English Northdown hops (whole)
            75 minutes (bittering)
2.5 HBU   (71 MBU) English Bramling Cross hops
            (whole)—75 minutes (bittering)
5   HBU   (142 MBU) English Fuggles hops (whole)—20
            minutes (flavor)
1   oz.   (28 g.) English Kent Goldings hops (whole)—
            steep in finished hot wort 2–5 minutes
            (aroma)

¼   tsp. Irish moss
¾   c.   corn sugar for priming in bottles. Use ⅓ cup
            corn sugar if priming a keg.

Wyeast 1275 Thames Valley Ale yeast is suggested. Use a
neutral, well-attenuating ale yeast that does not produce
excessive fruitiness. Diacetyl character is not desirable.

A single-step infusion mash is employed to mash the
grains. Add 7 quarts (6.65 l.) of 167-degree F (75 C) water
to the crushed grain, stir, stabilize and hold the temperature
at 150 degrees F (65.5 C) for 60 minutes.

After conversion, raise temperature to 167 degrees F (75
C), lauter and sparge with 4 gallons (15 l.) of 170-degree F
(77 C) water. Collect about 5.5 gallons (21 l.) of runoff, add
bittering hops and bring to a full and vigorous boil.

The total boil time will be 75 minutes. When 20 minutes
remain, add flavor hops. When 10 minutes remain, add Irish
moss. After a total wort boil of 75 minutes (reducing the
wort volume to just over 5 gallons), turn off the heat, add
aroma hops and let steep 2 to 5 minutes. Then separate or
strain out and sparge hops. Chill the wort to 65 to 70 degrees F (18–21 C) and direct into a sanitized fermenter. Aerate the cooled wort well. Add an active yeast culture and
ferment for 4 to 6 days in the primary. Then transfer into a
secondary fermenter, chill to 60 degrees F (15.5 C) if
possible.

When secondary aging is complete, prime with sugar, bottle

or keg. Let condition at temperatures above 60 degrees F (15.5 C) until clear and carbonated.

MALT-EXTRACT RECIPE AND PROCEDURE FOR RUDDLES BEST BITTER

4	lbs.	(1.8 kg.)	English extralight dried malt extract
½	lb.	(.23 kg.)	English crystal malt—90 Lovibond Cara-Pils
½	**lb.**	**(.23 kg.)**	**Total grains**

4	HBU	(113 MBU) English Northdown hops (whole) 60 minutes (bittering)
3	HBU	(85 MBU) English Bramling Cross hops (whole)—60 minutes (bittering)
5	HBU	(142 MBU) English Fuggles hops (whole)—20 minutes (flavor)
1	oz.	(28 g.) English Kent Goldings hops (whole)—steep in finished hot wort 2–5 minutes (aroma)

| ¼ | tsp. | Irish moss |
| ¾ | c. | corn sugar for priming in bottles. Use ⅓ cup corn sugar if priming a keg. |

Wyeast 1275 Thames Valley Ale yeast is suggested. Use a neutral, well-attenuating ale yeast that does not produce excessive fruitiness. Diacetyl character is not desirable.

Steep crushed crystal malt in 1½ gallons (5.7 l.) water at 150 degrees F (65.5 C) for 30 minutes. Strain and sparge with enough 170-degree F (76.5 C) water to finish with 2½ gallons (9.5 l.) specialty grain liquor. Add the dried malt extract and bittering hops and bring to a full and vigorous boil.

The total boil time will be 60 minutes. When 20 minutes remain, add flavor hops. When 10 minutes remain, add Irish moss. After a total wort boil of 60 minutes, turn off the heat, separate or strain out and sparge hops, and direct the hot wort into a sanitized fermenter to which 2 gallons (7.6 l.)

of cold water have been added. If necessary, add additional cold water to achieve a 5-gallon (19 l.) batch size. Chill the wort to 70 degrees F (21 C). Aerate the cooled wort well. Add an active yeast culture and ferment for 4 to 6 days in the primary. Then transfer into a secondary fermenter, chill to 60 degrees F (15.5 C) if possible.

When secondary aging is complete, prime with sugar, bottle or keg. Let condition at temperatures above 60 degrees F (15.5 C) until clear and carbonated.

Silver Cup Winner

Boddington's Bitter
Boddington's Brewery
Manchester, England

Also known as Boddington's Pub Ale Draught (with the "Draughtflow System"). Creamy, rich head is established with the release of nitrogen from a special system within the can. Rich color of light gold verging on amber. Slight English Kent hop aroma, but rather neutral overall. Mild carbonation. First flavor impression is light to medium-bodied accompanied by a smooth and excellent hop flavor and bitterness. Hop bitterness is the signature character of this beer; clean, punctual, soft and pleasant to the palate. Some fruitiness in the aroma. Malt character is not overstated in any one area. Good basic pale ale malt. Overall impression is smooth and velvety with a nice hop bite, both soft and memorable. Visually pleasing cling of foam on glass.

Estimated profile based on tasting
Color: 5–6 SRM (10–12 EBC)
Bittering Units: 27–32

Bronze Cup Winner

None

CATEGORY 4:
ENGLISH-STYLE (SPECIAL) BEST BITTER

Special Bitter is more robust than Ordinary Bitter. It has medium body and medium residual sweetness. In addition, Special Bitter has more hop character than Ordinary Bitter.

Original Gravity (°Plato):	1.038–1.045 (9.5–11 °Plato)
Apparent Extract–	
Final Gravity (°Plato):	1.006–1.012 (1.5–3 °Plato)
Alcohol by weight (volume):	3.3–3.8% (4.1–4.8%)
Bitterness (IBU):	28–46
Color SRM (EBC):	12–14 (30–35 EBC)

Gold Cup Winner

Delaney's Irish Ale
South China Brewing Co. Ltd.
Unit A1, 1/F, Vita Tower
29 Wong Chuk Hang
Aberdeen, Hong Kong
Brewmaster: Ted Miller
Established 1994
Production: 5,200 bbl. (6,100 hl.)

Edward L. Miller

A world-class English-style best bitter in Hong Kong? I suppose it's as likely as a world-class beer from your own homebrewery, and that's a sure bet. Established in 1994 by a group of investors from the United States and Hong Kong, the South

China Brewing Company is the first of what will be many small micro and pub breweries in Hong Kong in the coming years. Now the American penchant for introducing new ideas and ways of doing business has taken hold in Hong Kong and other parts of Asia. Handcrafted beers based on traditional styles and processes are emerging in the fastest-growing beer market in the world. With the American flair for entrepreneurship and the Hong Kong legacy of achievement, the small South China Brewing Company began brewing in June of 1995. It is an oasis in a land of transition. Beer enthusiasts are finding their products a welcome and fresh alternative to the typical light lagers available in the area. The brewery's best customers seem to be adventurous Americans who are pleasantly surprised to find authentic British-style ales, brewed to British standards and traditions.

Hong Kong became part of China in 1997. South China Brewing Company expects to make the transition and provide an example for future small brewery ventures. All the beers are brewed with an American-made 20-barrel brewhouse system. Besides award-winning Irish Ale, you'll find Crooked Island Ale (an American Pale Ale), Dragon's Back India Pale Ale, Stonecutters Lager (a Dortmund-style light lager), Signal 8 Stout, Breen' Brew ESB and seasonal brews available in bottles and on draft at popular beer bars throughout Hong Kong. The brewery anticipates exporting some of their products to the United States in 1997.

CHARACTER DESCRIPTION OF GOLD CUP–WINNING IRISH ALE

Appearance includes a slight but stylistically acceptable chill haze accenting the gentle amber-orange hue and the appealing fine and long-lasting bubbles. Delaney's Irish Ale has a complex and intense fruity aroma with hints of an applelike character. Though not immediately identifiable, there is a pleasing low level of diacetyl balancing the overall character of this best bitter. The initial perception of flavor impresses with a medium body and an immediate coincidence with the classic style descriptors. The gentle carbonation combines well with a soft but assertive hop-bitterness character; appropriate levels of bitterness and hop flavor emerge. The mild diacetyl also emerges in the flavor, rounding out the complex fruity character. Diacetyl flavor is less dominant than aroma. Artfully balanced mild pale ale malt flavor is sweetish but not caramel-like. In balance the diacetyl is more evident than caramel. Appropriately pleasant bitter aftertaste scores on the palate without being cloying. A slight musty character emerges as an afterthought, reminiscent of a beer aged in wood with a hint of tannic dryness in the palate. Overall, a remarkable best bitter in a bottle!

- **Recipe for 5 U.S. gallons (19 liters) Delaney's Irish Ale**
  Targets:
  Original Gravity: 1.050 (12.5)
  Final Gravity: 1.013 (3.3)
  Alcohol by volume: 5%
  Color: 13 SRM (26 EBC)
  Bittering Units: 36

ALL-GRAIN RECIPE AND PROCEDURE

8½	lbs. (3.6 kg.)	English 2-row pale malt
¼	lb. (0.11 kg.)	English Cara-Pils
½	lb. (0.23 kg.)	English crystal—80 Lovibond
9¼	**lbs. (4.2 kg.)**	**Total grains**

6	HBU	(170 MBU) American-grown Chinook hops (pellets)—60 minutes (bittering)
5	HBU	(142 MBU) American-grown Fuggles hops (pellets)—20 minutes (flavor)
½	oz.	(14 g.) American-grown Fuggles hops (pellets)—steep in finished wort 5 minutes (aroma)

1½	tsp.	gypsum
¼	tsp.	Irish moss
¾	c.	corn sugar for priming in bottles. Use ⅓ cup corn sugar if priming a keg.

Wyeast 1098 British Ale yeast (Whitbread origins)

Add ½ teaspoon of gypsum to the mash water and one teaspoon of gypsum to sparge water. A single-step infusion mash is employed to mash the grains. Add 9 quarts (8.6 l.) of 170-degree F (77 C) water to the crushed grain, stir, stabilize and hold the temperature at 155 degrees F (68 C) for 60 minutes.

After conversion, raise temperature to 167 degrees F (75 C), lauter and sparge with 4.5 gallons (17.1 l.) of 170-degree F (77 C) water. Collect about 5.5 to 5.75 gallons (21 to 22 l.) of runoff, add bittering hops and bring to a full and vigorous boil.

The total boil time will be 75 minutes. When 20 minutes remain, add flavor hops. When 10 minutes remain, add Irish moss. After a total wort boil of 75 minutes (reducing the wort volume to just over 5 gallons), turn off the heat, add aroma hops and let steep 2 to 5 minutes. Then separate or strain out and sparge hops. Chill the wort to 65 to 70 degrees F (18–21 C) and direct into a sanitized fermenter. Aerate the cooled wort well. Add an active yeast culture and ferment for 4 to 6 days in the primary. Then transfer into a secondary fermenter, chill to 60 degrees F (15.5 C) if possible.

When secondary aging is complete, prime with sugar, bottle or keg. Let condition at temperatures above 60 degrees F (15.5 C) until clear and carbonated.

MALT-EXTRACT RECIPE AND PROCEDURE FOR DELANEY'S IRISH ALE

5½	lbs.	(2.5 kg.)	English extralight dried malt extract
½	lb.	(0.22 kg.)	English crystal malt––90 Lovibond
6	**lbs.**	**(2.7 kg.)**	**Total grains**

7	HBU (198 MBU)	American-grown Chinook hops (pellets)—60 minutes (bittering)
6	HBU (170 MBU)	American-grown Fuggles hops (pellets)—20 minutes (flavor)
½	oz. (14 g.)	American-grown Fuggles hops (pellets)—steep in finished wort 5 minutes (aroma)

1½	tsp.	gypsum
¼	tsp.	Irish moss
¾	c.	corn sugar for priming in bottles. Use ⅓ cup corn sugar if priming a keg.

Wyeast 1098 British Ale yeast (Whitbread origins)

Steep crushed specialty grains in 1½ gallons (5.7 l.) water at 150 degrees F (65.5 C) for 30 minutes. Strain and sparge with enough 170-degree F (76.5 C) water to finish with a little over 2½ gallons (9.5 l.) specialty grain liquor. Add the gypsum, dried malt extract and bittering hops and bring to a full and vigorous boil.

The total boil time will be 60 minutes. When 20 minutes remain, add flavor hops. When 10 minutes remain, add Irish moss. After a total wort boil of 60 minutes, turn off the heat, separate or strain out and sparge hops, and direct the hot wort into a sanitized fermenter to which 2 gallons (7.6 l.) of cold water have been added. If necessary, add additional cold water to achieve a 5-gallon (19-l.) batch size. Chill the wort to 70 degrees F (21 C). Aerate the cooled wort well. Add an active yeast culture and ferment for 4 to 6 days in the primary. Then transfer into a secondary fermenter, chill to 60 degrees F (15.5 C) if possible and add aroma hops. Allow to age for two weeks or more.

When secondary aging is complete, prime with sugar, bottle or keg. Let condition at temperatures above 60 degrees F (15.5 C) until clear and carbonated.

Silver Cup Winner

Galena ESB
Sun Valley Brewing Co.
Hailey, Idaho, USA

Orange-hued light amber ale. Zesty character of American Galena hops softly stated. Soft caramel malt shadows the careful infusion of hops. Terrific malt base, almost suggestive of a smoky toasted character. Wonderful combination and balance with hop bitterness. Hop flavor is evident, but bitterness highlights the overall character. Not easily accomplished with the intensity of the quite pungent Galena hop. Balancing act well done. Medium-bodied and appropriately so. Clean aftertaste. Ale-like fruitiness is complex, contributing to the overall impression. The quite skillful balance of pungent and soft hop characters with the subtly expressed caramel, toffee or toasted malts likely made a positive impression on the judges. Without compromise, it is able to stand alone and exemplify the basic essentials of a great ESB.

Estimated profile based on tasting
Color: 6–8 SRM (12–16 EBC)
Bittering Units: 26–30

Bronze Cup Winner

None

CATEGORY 5:
ENGLISH-STYLE (EXTRA SPECIAL) STRONG BITTER

Extra Special Bitter possesses medium to strong hop qualities in aroma, flavor and bitterness. The residual malt sweetness of this richly flavored, full-bodied bitter is more pronounced than in other bitters.

Original Gravity (°Plato):	1.046–1.060 (11.5–15 °Plato)
Apparent Extract– Final Gravity (°Plato):	1.010–1.016 (2.5–4 °Plato)
Alcohol by weight (volume):	3.8–4.6% (4.8–5.8%)
Bitterness (IBU):	30–55
Color SRM (EBC):	12–14 (30–35 EBC)

Gold Cup Winner

Stoddard's ESB
Stoddard's Brewhouse and Eatery
111 S. Murphy Ave.
Sunnyvale, California, USA 94086
Brewmaster: Bob Stoddard
Head Brewer: Mike Gray
Established 1993
Production: 2,000 bbl. (2,300 hl.)

Stoddard's Brewhouse and Eatery was founded and designed by Bob Stoddard in 1993. The site consists of a 2,000-square-foot brewhouse, 9,000-square-foot restaurant and a 2,000-square-foot outdoor beer garden. All their beers are single-infused mashes currently produced in their 20-barrel brewhouse.

Bob Stoddard, Brewmaster

Brewmaster Bob Stoddard and head brewer Mike Gray brew with the intensity of overachieving homebrewers. As a homebrewer yourself, you can appreciate and anticipate the results: production of several winning brews. Every beer they entered in the 1996 World Beer Cup was honored with an award. Stoddard's ESB and the Stoddard's Kölsch took Gold Cups while Stoddard's Porter won a Bronze Cup. Each of these beers also has won medals at past Great American Beer Festivals. A phenomenal and impressive accomplishment. Other winning Stoddard beers from other competitions include Stoddard's Pale Ale, Stoddard's Kristal Weizen, Stoddard's Helles Bock, Stoddard's Oatmeal Stout and seasonal beers.

CHARACTER DESCRIPTION OF GOLD CUP–WINNING STODDARD'S ESB
Tawny orange-brown color with amber hue. Crystal-clear even at chilled temperatures. An applelike fruitiness and alcoholic aroma combine to offer a first impression. Though hop aroma is initially absent, it does emerge and linger in the background as a sweetish, floral, spicy aroma vaguely reminiscent of a Kent Goldings–type hop, but not quite. Flavor impression is at first characteristic of dry, dark caramel followed with a subtle and refreshing roasted malt (black

malt or even roasted barley) astringency. Suggested dryness is balanced with a pleasant medium body. Not particularly sweet malty or sweet caramel-like. Bitterness is a harsh bitterness reflective of a higher alpha hop, but not overdone. Aftertaste is generally bitter with a tingling alcohol sensation in the mouth feel. Overall balance is not particularly complex, with clean neutral fermentation character. No diacetyl is perceived.

- **Recipe for 5 U.S. gallons (19 liters) Stoddard's ESB**
  Targets:
  Original Gravity: 1.050 (12.5)
  Final Gravity: 1.014 (3.5)
  Alcohol by volume: 5%
  Color: 14 SRM (28 EBC)
  Bittering Units: 32

ALL-GRAIN RECIPE AND PROCEDURE

8½	lbs.	(3.9 kg.)	American 2-row Klages pale malt
0.05	lb.	(23 g.)	roasted barley
½	lb.	(0.23 kg.)	English crystal malt—60 Lovibond
½	lb.	(0.23 kg.)	American wheat malt
9½	**lbs.**	**(4.3 kg.)**	**Total grains**

6.3 HBU (179 MBU) American Cascade hops (pellets)—75 minutes (bittering)

1.3 HBU (37 MBU) American Fuggles hops (pellets)—30 minutes (flavor)

¼ oz. (7 g.) American Cascade hops (pellets)—steep in finished boiled wort for 2 to 3 minutes (aroma)

¼ tsp. Irish moss

¾ c. corn sugar for priming in bottles. Use ⅓ cup corn sugar if priming a keg.

Wyeast 1056 American Ale yeast is suggested.

A single-step infusion mash is employed to mash the grains. Add 9.5 quarts (9 l.) of 167-degree F (75 C) water to the crushed grain, stir, stabilize and hold the temperature at 152 degrees F (67 C) for 60 minutes.

After conversion, raise temperature to 167 degrees F (75 C), lauter and sparge with 4.5 gallons (17.1 l.) of 170-degree F (77 C) water. Collect about 5.5 to 5.75 gallons (21 to 22 l.) of runoff, add bittering hops and bring to a full and vigorous boil.

The total boil time will be 75 minutes. When 30 minutes remain, add flavor hops. When 10 minutes remain, add Irish moss. After a total wort boil of 75 minutes (reducing the wort volume to just over 5 gallons), turn off the heat, add aroma hops and let steep 2 to 5 minutes. Then separate or strain out and sparge hops. Chill the wort to 65 to 70 degrees F (18–21 C) and direct into a sanitized fermenter. Aerate the cooled wort well. Add an active yeast culture and ferment for 4 to 6 days in the primary. Then transfer into a secondary fermenter, chill to 55 to 60 degrees F (13–15.5 C) if possible.

When secondary aging is complete, prime with sugar, bottle or keg. Let condition at temperatures above 60 degrees F (15.5 C) until clear and carbonated.

MALT-EXTRACT RECIPE AND PROCEDURE FOR STODDARD'S ESB

5	lbs.	(2.3 kg.)	English extralight dried malt extract
1	lb.	(0.45 kg.)	American wheat malt extract syrup (50% wheat)
0.05	lb.	(23 g.)	roasted barley
½	lb.	(0.23 kg.)	English crystal malt—60 Lovibond
½	**lb.**	**(0.23 kg.)**	**Total grains**

8	HBU	(227 MBU)	American Cascade hops (pellets)—60 minutes (bittering)
1.5	HBU	(43 MBU)	American Fuggles hops (pellets)—30 minutes (flavor)
¼	oz.	(7 g.)	American Cascade hops (pellets)—steep in finished boiled wort for 2 to 3 minutes (aroma)

¼   tsp.   Irish moss
¾   c.   corn sugar for priming in bottles. Use ⅓ cup
             corn sugar if priming a keg.
Wyeast 1056 American Ale yeast is suggested.

Steep crushed specialty grains in 1½ gallons (5.7 l.) water
at 150 degrees F (65.5 C) for 30 minutes. Strain and sparge
with enough 170-degree F (76.5 C) water to finish with a
little over 2½ gallons (9.5 l.) specialty grain liquor. Add the
dried and syrup malt extract and bittering hops and bring
to a full and vigorous boil.

The total boil time will be 60 minutes. When 30 minutes
remain, add flavor hops and Irish moss. After a total wort
boil of 60 minutes, turn off the heat, add aroma hops and
let steep 2 to 5 minutes. Then separate or strain out and
sparge hops, and direct the hot wort into a sanitized fermenter to which 2 gallons (7.6 l.) of cold water have been
added. If necessary, add additional cold water to achieve a
5-gallon (19 l.) batch size. Chill the wort to 70 degrees F
(21 C). Aerate the cooled wort well. Add an active yeast
culture and ferment for 4 to 6 days in the primary. Then
transfer into a secondary fermenter, chill to 55 to 60 degrees
F (13–15.5 C) if possible and add aroma hops. Allow to age
for two weeks or more.

When secondary aging is complete, prime with sugar, bottle or keg. Let condition at temperatures above 60 degrees
F (15.5 C) until clear and carbonated.

Silver Cup Winner

Extra Special Bitter Ale
Bison Brewing Co.
Berkeley, California, USA

Substantially amber with orange hue. Extraordinarily notable head and foam retention, unusual for most American

microbrews. Bottle conditioning surely contributes to this effect. Aroma reminiscent of toffee and earthy hop character, neither particularly obsessive, though stated well enough to arouse one's curiosity. With second and third whiffs there begins to emerge a very subtle yet alluring smoky aroma. Medium to full-bodied texture and flavor are extraordinarily smooth and notable characters. Hop flavor as well as smooth and gently soft bitterness are reminiscent of British bitter. Malt is well stated but not overly sweet. The smoothness and softness bottle conditioning can lend to the overall character is evident in this perfectly carbonated ale.

Estimated profile based on tasting
Color: 10–12 SRM (20–24 EBC)
Bittering Units: 37–40

Bronze Cup Winner

Rikenjaks ESB
Rikenjaks Brewing Co. Inc.
Lafayette, Louisiana, USA

Amber color with orange hue. A robust aroma greets your indulgence with a deep grainy texture and a mild but evident hop character. Fruitiness is apparent, but center stage is taken by malt and hops. Rich caramel malt character without excessive toffee or roast notes expresses itself both in flavor and aroma. Bitterness is substantial given the medium-bodied balance of this brew from Louisiana. Malt character follows the strong impact of bitterness. Malt sweetness begins the flavor journey, followed by hop bitterness and then the mellowing character of caramel malt. But then, just as you believe you've got the flavor all figured out, hop bitterness reemerges as a final statement, just as a good ESB should. It lingers long enough to make impressions with bitterness and a great Kent Goldings hop flavor.

(The label indicates the use of Kent Goldings along with nine different malts.)

Estimated profile based on tasting
Original Gravity: 1.054 as labeled
Final Gravity: 1.009 as labeled
Color: 10–12 SRM (20–24 EBC)
Bittering Units: 45–47

CATEGORY 6:
SCOTTISH-STYLE ALE

Characterized by a rounded flavor profile, Scottish Ales are malty, caramel-like, soft and chewy. Hop rates are low. Yeast characters such as diacetyl (butterscotch) and sulfuriness are acceptable at very low levels. Scottish Ales range from golden amber to deep brown in color and may possess a faint smoky character. Bottled versions of this traditional draft beer may contain higher amounts of carbon dioxide than is typical for draft versions. Chill haze is acceptable at low temperatures.

A. SUBCATEGORY: SCOTTISH LIGHT ALE

Scottish Light is the mildest form of this ale. Little bitterness is perceived. It is light-bodied with very low hop bitterness. No hop flavor or aroma is perceived. Chill haze is acceptable at low temperatures.

Original Gravity (°Plato):	1.030–1.035 (7.5–9 °Plato)
Apparent Extract–	
Final Gravity (°Plato):	1.006–1.012 (1.5–3 °Plato)
Alcohol by weight (volume):	2.2–2.8% (2.83–5%)

Bitterness (IBU):	9–20
Color SRM (EBC):	8–17 (16–40 EBC)

B. SUBCATEGORY: SCOTTISH HEAVY ALE

Scottish Heavy Ale is moderate in strength and dominated by a smooth, sweet maltiness that is balanced with low but perceptible hop bitterness. It has medium body. Fruity esters are very low if evident. Chill haze is acceptable at low temperatures.

Original Gravity (°Plato):	1.035–1.040 (9–10 °Plato)
Apparent Extract– Final Gravity (°Plato):	1.0010–1.014 (2..5–3.5 °Plato)
Alcohol by weight (volume):	2.8–3.2% (3.5–4%)
Bitterness (IBU):	12–20
Color SRM (EBC):	10–19 (20–75 EBC)

C. SUBCATEGORY: SCOTTISH EXPORT ALE

Scottish Export Ale is sweet, caramel-like and malty. Its bitterness is perceived as low to medium. It has medium body. Fruity ester character may be apparent. Chill haze is acceptable at low temperatures.

Original Gravity (°Plato):	1.040–1.050 (10–12.5 °Plato)
Apparent Extract– Final Gravity (°Plato):	1.010–1.018 (2.5–4.5 °Plato)
Alcohol by weight (volume):	3.2–3.6% (4.0–4.5%)
Bitterness (IBU):	15–25
Color SRM (EBC):	10–19 (20–75 EBC)

Gold Cup Winner

Samuel Adams Scotch Ale
Philadelphia Brewing Company/Samuel Adams Brewhouse
1516 Sansom St.
Philadelphia, Pennsylvania, USA 19102
Head Brewer: Mike Carota
Established 1989
Production: 1,000 bbl. (1,200 hl.)

Part of the family of Boston Beer Company's brewing ventures, this small brewpub located in downtown Philadelphia features an assortment of ales and lagers year-round. The Boston Beer Company was founded in 1984 and continues to offer the widest assortment of beer styles of any brewer in the world. These products are brewed at several locations around the country, most in large breweries having packaging equipment able to handle the 1 million–plus barrels of

Jim Koch, Boston Beer Company

beer produced per year. The brewhouse in Philadelphia, solely owned and operated by the Boston Beer Company, produces beer under the guidance of master brewer Jim Koch, brewing beer in the craft tradition of the smallest brewers in America. You'll find versions of porters, stouts, Oktoberfests, pale ales, light lagers and more on the second floor at 1516 Sansom Street.

CHARACTER DESCRIPTION OF GOLD CUP–WINNING SAMUEL ADAMS
SCOTCH ALE

Color is a deep ruby red-brown. Full malt aroma with suggestion of smoked malt rounding the rich character of the aroma. Extraordinarily well-balanced aroma with hop character appropriately absent. Caramel and roast malts combined with peat-smoked malt. The smoked malt is done very subtly, not competing as the primary character. Full-bodied texture with flavor bearing the same character that is evident in the aroma; full malt flavor. Alcohol strength is evident but deceivingly subtle due to the richness of body and fullness of malt. Hop character is all expressed as medium bitterness, quite adequate to balance the fullness of malt. Aftertaste is clean; malt lingers and gently gives way to the emergence of a pleasant bitterness and smoky memory. Excellent rendition of this style.

• **Recipe for 5 U.S. gallons (19 liters) Samuel Adams Scotch Ale**
Targets:
Original Gravity: 1.060 (15)
Final Gravity: 1.022 (5.4)
Alcohol by volume: 5.5%
Color: 19 SRM (38 EBC)
Bittering Units: 30

ALL-GRAIN RECIPE AND PROCEDURE
10¼ lbs. (4.7 kg.)   English 2-row pale malt
½   lb. (0.23 kg.)   American (aromatic) Victory malt

0.10 lb.   (50 g.)   Scottish peat-smoked malt
0.05 lb.   (23 g.)   American chocolate malt
½   lb.   (0.23 kg.)   American caramel malt—60 Lovibond
11.4 lbs. (5.2 kg.)   Total grains

6.3   HBU   (179 MBU) English Fuggles hops (pellets)—90
            minutes (bittering)
1.3   HBU   (37 MBU) English Kent Goldings (pellets)—30
            minutes (flavor)

¼   tsp.   Irish moss
¾   c.   corn sugar for priming in bottles. Use ⅓ cup
           corn sugar if priming a keg.
Wyeast 1728 Scottish Ale yeast can be recommended.

A single-step infusion mash is employed to mash the grains. Add 11.5 quarts (11 l.) of 172-degree F (78 C) water to the crushed grain, stir, stabilize and hold the temperature at 156 degrees F (69 C) for 60 minutes.

After conversion, raise temperature to 167 degrees F (75 C), lauter and sparge with 4 gallons (15 l.) of 170-degree F (77 C) water. Collect about 6 to 6.5 gallons (25 l.) of runoff, add bittering hops and bring to a full and vigorous boil.

The total boil time will be 90 minutes in order to evaporate the wort to about 5.25 gallons. When 30 minutes remain, add flavor hops. When 10 minutes remain, add Irish moss. After a total wort boil of 90 minutes, turn off the heat, then separate or strain out and sparge hops. Chill the wort to 70 degrees F (21 C) and direct into a sanitized fermenter. Aerate the cooled wort well. Add an active yeast culture and ferment for 4 to 6 days in the primary. Then transfer into a secondary fermenter, chill to 55 to 60 degrees F (13–15.5 C) if possible.

When secondary aging is complete, prime with sugar, bottle or keg. Let condition at temperatures above 60 degrees F (15.5 C) until clear and carbonated.

MASH-EXTRACT RECIPE AND PROCEDURE FOR SAMUEL ADAMS SCOTCH ALE

3¼ lbs.	(1.5 kg.)	British amber dried malt extract
1¼ lbs.	(0.57 kg.)	British light dried malt extract
3 lbs.	(1.4 kg.)	English 2-row pale malt
½ lb.	(0.23 kg.)	American (aromatic) Victory malt
0.10 lb.	(50 g.)	Scottish peat-smoked malt
0.05 lb.	(23 g.)	American chocolate malt
½ lb.	(0.23 kg.)	American caramel malt—60 Lovibond
4.15 lbs.	**(1.9 kg.)**	**Total grains**

7.5 HBU	(213 MBU)	English Fuggles hops (pellets)—75 minutes (bittering)
1.3 HBU	(37 MBU)	English Kent Goldings (pellets)—30 minutes (flavor)

¼ tsp.		Irish moss
¾ c.		corn sugar for priming in bottles. Use ⅓ cup corn sugar if priming a keg.

Wyeast 1728 Scottish Ale yeast can be recommended.

A single-step infusion mash is employed to mash the grains. Add 4.15 quarts (4 l.) of 172-degree F (78 C) water to the crushed grain, stir, stabilize and hold the temperature at 156 degrees F (69 C) for 60 minutes.

After conversion, raise temperature to 167 degrees F (75 C), lauter and sparge with 2 gallons (7.6 l.) of 170-degree F (77 C) water. Collect about 30 gallons (11.5 l.) of runoff. Add malt extract and bittering hops and bring to a full and vigorous boil.

The total boil time will be 75 minutes. When 30 minutes remain, add flavor hops. When 10 minutes remain, add Irish moss. After a total wort boil of 75 minutes, turn off the heat, separate or strain out and sparge hops, and direct the hot wort into a sanitized fermenter to which 2 gallons (7.6 l.) of cold water have been added. If necessary, add additional cold water to achieve a 5-gallon (19 l.) batch size.

Chill the wort to 70 degrees F (21 C). Aerate the cooled wort well. Add an active yeast culture and ferment for 4 to 6 days in the primary. Then transfer into a secondary fermenter, chill to 55 to 60 degrees F (13–15.5 C) if possible. Allow to age for two weeks or more.

When secondary aging is complete, prime with sugar, bottle or keg. Let condition at temperatures above 60 degrees F (15.5 C) until clear and carbonated.

Silver Cup Winner

Laughing Lab Scottish Ale
Bristol Brewing
Colorado Springs, Colorado, USA

Deep reddish color with hints of dark orange or amber. Great full malt aroma complemented with ale fruitiness. Hints of American hops in aroma: spicy, citrusy and Cascade-like. Medium-bodied ale with full malt character, finishing with a clean aftertaste. Well balanced with obvious bitterness, though only fulfilling the role of balancing malt and fruity characters in flavor. Fruitiness is plum or wild cherrylike, and expresses itself from start to finish. Aftertaste begins with a slight caramel-toffee malt character and zesty fruitiness, fades to a brief visit with bitterness, then trails off to another encounter with your lips . . . and another and another. Roast malt characters do not emerge; rather, a pleasant soft, dark caramel malt character subtly dominates in this arena. A great example of this style.

Brewery formulation uses crystal malts, chocolate, Cara-Pils and Vienna malt with Chinook, Willamette and Mt. Hood for bitterness, flavor and aroma.

Estimated profile based on tasting
Original Gravity: 1.054 (13.5) indicated by the brewery
Final Gravity: 1.015 (3.8) indicated by the brewery

Color: 19–21 SRM (38–42 EBC)
Bittering Units: 30–35 (estimated 22 indicated by the brewery)

Bronze Cup Winner

Right Field Red Ale
Sandlot Brewery at Coors Field
Denver, Colorado, USA

Deep ruby red amber ale tending toward an overall brown color. Notably low carbonation. No hop aroma evident. Lots of alelike fruitiness. Clean, malty and dry. Hop bitterness is low but enough to balance the high alcohol content and full malt flavor. Fruity flavor is complex and slightly acidic, with slight bitterness following in the aftertaste. Despite the bigness of several characters, beer finishes quite dry, enhancing its drinkability. As one dwells with Right Field Red Ale and it warms, a slight toasted malt character emerges, suggesting this beer is better served at 50 degrees F (10 C) rather than 40 F (4.5 C).

Estimated profile based on tasting
Color: 13–14 SRM (26–28 EBC)
Bittering Units: 35–36

CATEGORY 7:
ENGLISH-STYLE MILD ALE

A. SUBCATEGORY: ENGLISH-STYLE LIGHT MILD ALE

English-style Light Mild Ales range from light amber to light brown in color. Malty sweet tones dominate the flavor profile, with little hop bitterness or flavor. Hop aroma can be light. Very low diacetyl flavors may be appropriate in this low-alcohol beer. Fruity ester level is very low. Chill haze is allowable at cold temperatures.

Original Gravity (°Plato):   1.030–1.038 (7.5–9.5 °Plato)
Apparent Extract–
Final Gravity (°Plato):   1.004–1.008 (1–2 °Plato)
Alcohol by weight (volume):   2.7–3.2% (3.2–4.0%)
Bitterness (IBU):   10–24
Color SRM (EBC):   8–17 (16–40 EBC)

B. SUBCATEGORY: ENGLISH-STYLE DARK MILD ALE

English-style Dark Mild Ales range from deep copper to dark brown (often with a red tint) in color. Malty sweet, caramel, licorice and roast malt tones dominate the flavor and aroma profile, with very little hop flavor or aroma. Very low diacetyl flavors may be appropriate in this low-alcohol beer. Fruity ester level is very low.

Original Gravity (°Plato):   1.030–1.038 (7.5–9.5 °Plato)
Apparent Extract–
Final Gravity (°Plato):   1.004–1.008 (1–2 °Plato)
Alcohol by weight (volume):   2.7–3.2% (3.2–4.0%)
Bitterness (IBU):   10–24
Color SRM (EBC):   17–34 (40–135 EBC)

Gold Cup Winner

Seabright Session Ale
Seabright Brewery
519 Seabright Ave.
Santa Cruz, California, USA 95062
Brewmaster: Will Turner
Established 1988
Production: 1,200 bbl. (1,400 hl.)

Three blocks from the ocean and minutes from the Santa Cruz Yacht Harbor and boardwalk, the Seabright Brewery brings food, beer and setting together in a manner that makes a visit a great episode in the search for life's pleasures. The origins of this multi-award-winning brewery have all the features of a classic homebrewed beginning. It was award-winning homebrew that initiated Charlie Meehan's journey from homebrewer to owner of the 7-barrel brewhouse with partner Keith Cranmer. While their Session Ale won in the World Beer Cup, other beers have won honors in the Great American Beer Festival since 1991. These include Seabright Amber Ale (Gold Medal at the 1991 GABF, in the American Amber Ale category), Oatmeal Stout (Gold Medal at the 1992 GABF, in the Sweet Stout category), Banty Rooster IPA (Gold Medal at the 1991 GABF, in the India Pale Ale category), and Pleasure Point Porter (Silver Medal at the 1992 GABF, in the Porter category). Pelican Pale Ale and a continually changing variety of specialty beers such as English-style bitter, wheat beer, bock and barley wine also can be found on tap near the waterfront at the Seabright Brewery.

CHARACTER DESCRIPTION OF GOLD CUP–WINNING SEABRIGHT SESSION ALE
Deep reddish brown, with orange reddish hue. Strong malty sweet, toasted character predominates in the aroma, almost suggesting the character of diacetyl, but this sugges-

tion can be dismissed by an imagined healthy dose of aromatic caramel or crystal. Aroma is suggestive of the caramel lollipop called Sugar Daddy. The first impression is of a light-bodied dry ale with high drinkability. Though intense, the aromatic maltiness does not overpower the flavor. Wonderfully balanced with suggestions of roast but not burnt-malt bitterness. Roast-malt dryness and subtle cocoa character are evident. Well attenuated and very quenching, with lightness in the mouth feel. Hop bitterness is in perfect balance, not suggesting bitterness but evident and providing the counterpoint to the malt character. Aftertaste is clean. Diacetyl is absent in the flavor. As its name suggests, this is a beer that is not tiring. Adequately carbonated, more so than a typical English mild, but very refreshing.

- ### Recipe for 5 U.S. gallons (19 liters) Seabright Session Ale
  Targets:
  Original Gravity: 1.036 (9)
  Final Gravity: 1.006 (1.5)
  Alcohol by volume: 3.75%
  Color: 23–25 SRM (46–50 EBC)
  Bittering Units: 25

ALL-GRAIN RECIPE AND PROCEDURE

2¾	lbs.	(1.2 kg.)	American 2-row Klages pale malt
2¾	lbs.	(1.2 kg.)	English 2-row pale malt
1	lb.	(0.45 kg.)	American caramel malt—80 Lovibond
¼	lb.	(114 g.)	American wheat malt
¼	lb.	(114 g.)	American Victory or other aromatic malt
0.1	lb.	(45 g.)	American chocolate malt
0.1	lb.	(45 g.)	American black malt
6.2	**lbs.**	**(2.8 kg.)**	**Total grains**

4   HBU   (113 MBU) American Fuggles hops (pellets)—
          60 minutes (bittering)
1.25 HBU   (35 MBU) American Fuggles hops (whole)—30
          minutes (bittering)
1.25 HBU   (35 MBU) American Goldings hops (whole)—
          30 minutes (bittering)
¼   oz.   (7 g.) American Fuggles hops (pellets)—steep
          in finished boiled wort for 2 to 3 minutes
          (aroma)

¼   tsp. Irish moss
¾   c.   corn sugar for priming in bottles. Use ⅓ cup
          corn sugar if priming a keg.
Wyeast 1056 American Ale yeast

A step infusion mash is employed to mash the grains.
Add 6 quarts (5.7 l.) of 138-degree F (58 C) water to the
crushed grain, stir, stabilize and hold the temperature at
128 degrees F (53 C) for 30 minutes. Add 3 quarts (2.9 l.)
of boiling water. Add heat to bring temperature up to 150
degrees F (65.5 C). Hold for about 60 minutes.

After conversion, raise temperature to 167 degrees F (75
C), lauter and sparge with 4 gallons (15.2 l.) of 170-degree
F (77 C) water. Collect about 5.5 gallons (21 l.) of runoff,
add bittering hops and bring to a full and vigorous boil.

The total boil time will be 60 minutes. When 30 minutes
remain, add flavor hops. When 10 minutes remain, add Irish
moss. After a total wort boil of 60 minutes, turn off the
heat, add aroma hops and let steep 2 to 5 minutes. Then
separate or strain out and sparge hops. Chill the wort to 70
degrees F (21 C) and direct into a sanitized fermenter. Aerate the cooled wort well. Add an active yeast culture and
ferment for 4 to 6 days in the primary. Then transfer into a
secondary fermenter, chill to 55 to 60 degrees F (13–15.5 C)
and allow to age for two weeks or more.

When secondary aging is complete, prime with sugar, bot-

tle or keg. Let condition at temperatures above 60 degrees F (15.5 C) until clear and carbonated.

MASH-EXTRACT RECIPE AND PROCEDURE FOR SEABRIGHT SESSION ALE

2	lbs.	(0.9 kg.)	English extralight dried malt extract
2	lbs.	(0.9 kg.)	English 2-row pale malt
1	lb.	(0.45 kg.)	American caramel malt—80 Lovibond
¼	lb.	(114 g.)	American wheat malt
¼	lb.	(114g.)	American Victory or other aromatic malt
0.1	lb.	(45 g.)	American chocolate malt
0.1	lb.	(45 g.)	American black malt
3.7	**lbs.**	**(1.7 kg.)**	**Total grains**

5	HBU	(142 MBU) American Fuggles hops (pellets)—60 minutes (bittering)
1.25	HBU	(35 MBU) American Fuggles hops (whole)—30 minutes (bittering)
1.25	HBU	(35 MBU) American Goldings hops (whole)—30 minutes (bittering)
¼	oz.	(7 g.) American Fuggles hops (pellets)—steep in finished boiled wort for 2 to 3 minutes (aroma)

¼	tsp.	Irish moss
¾	c.	corn sugar for priming in bottles. Use ⅓ cup corn sugar if priming a keg.

Wyeast 1056 American Ale yeast

A step infusion mash is employed to mash the grains. Add 3.75 quarts (3.5 l.) of 138-degree F (58 C) water to the crushed grain, stir, stabilize and hold the temperature at 128 degrees F (53 C) for 30 minutes. Add 2 quarts (2.9 l.) of boiling water. Add heat to bring temperature up to 150 degrees F (65.5 C). Hold for about 60 minutes.

After conversion, raise temperature to 167 degrees F (75 C), lauter and sparge with 2 gallons (7.6 l.) of 170-degree F (77 C) water. Collect about 2.5–2.75 gallons (9.5–10.5 l.) of runoff. Add malt extract and bittering hops and bring to a full and vigorous boil.

The total boil time will be 60 minutes. When 30 minutes remain, add flavor hops. When 10 minutes remain, add Irish moss. After a total wort boil of 60 minutes, turn off the heat, add aroma hops and let steep 2 to 5 minutes. Then separate or strain out and sparge hops, and direct the hot wort into a sanitized fermenter to which 2 gallons (7.6 l.) of cold water have been added. If necessary, add additional cold water to achieve a 5-gallon (19 l.) batch size. Chill the wort to 70 degrees F (21 C). Aerate the cooled wort well. Add an active yeast culture and ferment for 4 to 6 days in the primary. Then transfer into a secondary fermenter, chill to 55 to 60 degrees F (13–15.5 C) and allow to age for two weeks or more.

When secondary aging is complete, prime with sugar, bottle or keg. Let condition at temperatures above 60 degrees F (15.5 C) until clear and carbonated.

Silver Cup Winner

None

Bronze Cup Winner

None

CATEGORY 8:
ENGLISH-STYLE BROWN ALE

English-style Brown Ales range from deep copper to brown in color. They have a medium body. Dry to sweet maltiness dominates, with very little hop flavor or aroma. Fruity ester flavors are appropriate. Diacetyl should be very low if evident. Chill haze is allowable at cold temperatures.

Original Gravity (°Plato):	1.040–1.050 (10–12.5 °Plato)
Apparent Extract– Final Gravity (°Plato):	1.008–1.014 (2–3.5 °Plato)
Alcohol by weight (volume):	3.3–4.7% (4–5.5%)
Bitterness (IBU):	15–25
Color SRM (EBC):	15–22 (35–90 EBC)

Gold Cup Winner

Redwood Coast Brown Ale
Redwood Coast Brewing Company
1051 Pacific Marina
Alameda, California, USA 94501
Brewmaster: Dr. Ronald Manabe
Established 1987
Production: 4,500 bbl. (5,300 hl.)

The Redwood Coast Brewing Company is no stranger to making award-winning beer in the name of the founding Tied House Café and Brewery, a brewery-restaurant in downtown Mountain View, California. The company's founder and first master brewer, Cheuck Tom, was once the chief chemist and then brewmaster at San Miguel Brewing in Hong Kong for over thirty years. He then joined Anheuser-Busch and became assistant brewmaster at the Los Angeles brewery. After his retirement Mr. Tom founded the first Tied House Café and Brewery with his creative energy and dedication to producing a variety of excellent European and pre-Prohibition–style American beers. Mr. Tom passed away in 1993. Dr. Andreas Heller continued the tradition of maintaining and developing beers for the three Tied House Café and

Ronald Manabe

(*Descending*) *John Sullivan, Sylvestre Fortiz and Scott Phillips, Brewers*

Breweries. (Others have opened in Alameda and San Jose, California.) The packaged products now produced for distribution to other restaurants and retail outlets are sold under the name of The Redwood Coast Brewing Company. Dr. Ronald Manabe, a Siebel graduate and a member of the original founding team, took over brewmaster responsibilities in March of 1996 and continues to uphold the quality tradition of the little brewery that *did*.

The brewery's products include award-winning Wheat (1990 GABF Bronze Medal in the American Wheat category), Amber (1988 GABF Bronze Medal and 1989 Silver Medal in the Brown Ale category), Dry (1991 GABF Silver Medal in the Blonde/Golden Ale category), Dark (1991 GABF Gold Medal in the Brown Ale category), and Passion Pale (1992 GABF Gold Medal and 1993 GABF Silver Medal in the Fruit Beer category). These products are now available in 5- and 15.5-gallon kegs as Alpine Gold, Cascade Amber, Redwood Brown, New World Wheat, Redwood Coast Oktoberfest, Raspberry Wheat and Passion Pale.

CHARACTER DESCRIPTION OF GOLD CUP–WINNING REDWOOD COAST BROWN ALE

Deep chestnut brown with reddish ruby hue. Aroma has a complex light fruitiness (reminiscent of plums and strawberries) and is authentically alelike with a strong suggestion of cocoa and a light but rich aroma of caramel. Flavor at first impression is very smooth and not astringent. Caramel and roasted-dark-malt characters are delicate and sweet. Hop flavor is neutral and reminiscent of American-grown varieties. Hop aroma and flavor are memorable. Extremely drinkable, but with a touch more bitterness than most classic English browns. Body is medium with an aftertaste that finishes toward the dry and slightly bitter end. An admirably difficult balance seems to have been achieved: sweet but without an overwhelming sense of caramel, sweet but not full-bodied, rather dryish. Clean with no diacetyl.

- ### *Recipe for 5 U.S. gallons (19 liters) Redwood Coast Brown Ale*
  Targets:
  Original Gravity: 1.052 (13)
  Final Gravity: 1.014 (3.5)
  Alcohol by volume: 5%
  Color: 22 SRM (44 EBC)
  Bittering Units: 30

ALL-GRAIN RECIPE AND PROCEDURE

4¼	lbs.	(1.9 kg.) American 2-row Klages pale malt
4¼	lbs.	(1.9 kg.) English 2-row pale malt
½	lb.	(0.23 kg.) American caramel malt—70/80 Lovibond
½	lb.	(0.23 kg.) American caramel malt—135/165 Lovibond
0.1	lb.	(45 g.) American chocolate malt
0.1	lb.	(45 g.) American black malt
9.7	**lbs.**	**(3.2 kg.) Total grains**

1.5   HBU   (43 MBU) Cascade hops (pellets)—75 minutes (bittering)

5   HBU   (142 MBU) American Willamette hops (pellets)—75 minutes (bittering)

½   oz.   (14 g.) European Saaz hops (pellets)   steep in finished boiled wort for 2 to 3 minutes (aroma)

¼   tsp.   Irish moss

¾   c.   corn sugar for priming in bottles. Use ⅓ cup corn sugar if priming a keg.

Wyeast 1098 British Ale yeast (Whitbread origin)

A step infusion mash is employed to mash the grains. Add 10 quarts (9.5 l.) of 138-degree F (58 C) water to the crushed grain, stir, stabilize and hold the temperature at 128 degrees F (53 C) for 30 minutes. Add 5 quarts (4.75 l.) of boiling water, add heat to bring temperature up to 156 degrees F (69 C), and hold for about 60 minutes.

After conversion, raise temperature to 167 degrees F (75 C), lauter and sparge with 4 gallons (15 l.) of 170-degree F (77 C) water. Collect about 6 gallons (23 l.) of runoff, add bittering hops and bring to a full and vigorous boil.

The total boil time will be 75 minutes. When 10 minutes remain, add Irish moss. After a total wort boil of 75 minutes (reducing the wort volume to just over 5 gallons), turn off the heat, add aroma hops and let steep 2 to 5 minutes. Then separate or strain out and sparge hops. Chill the wort to 65 to 70 degrees F (18–21 C) and direct into a sanitized fermenter. Aerate the cooled wort well. Add an active yeast culture and ferment for 4 to 6 days in the primary. Then transfer into a secondary fermenter, chill to 55 to 60 degrees F (13–15.5 C) if possible and add aroma hops. Allow to age for two weeks or more.

When secondary aging is complete, prime with sugar, bottle or keg. Let condition at temperatures above 60 degrees F (15.5 C) until clear and carbonated.

MALT-EXTRACT RECIPE AND PROCEDURE FOR REDWOOD COAST BROWN ALE

5¼	lbs.	(2.4 kg.)	English light dried malt extract
½	lb.	(0.23 kg.)	American caramel malt—70/80 Lovibond
½	lb.	(0.23 kg.)	American caramel malt—135/165 Lovibond
0.1	lb.	(45 g.)	American chocolate malt
0.1	lb.	(45 g.)	American black malt
1.2	**lbs.**	**(0.5 kg.)**	**Total grains**

2	HBU	(56 MBU) Cascade hops (pellets)—60 minutes (bittering)
6	HBU	(170 MBU) American Willamette hops (pellets)—60 minutes (bittering)
½	oz.	(14 g.) European Saaz hops (pellets)—steep in finished boiled wort for 2 to 3 minutes (aroma)

¼	tsp.	Irish moss
¾	c.	corn sugar for priming in bottles. Use ⅓ cup corn sugar if priming a keg.

Wyeast 1098 British Ale yeast (Whitbread origin)

Steep crushed specialty grains in 1½ gallons (5.7 l.) water at 150 degrees F (65.5 C) for 30 minutes. Strain and sparge with enough 170-degree F (76.5 C) water to finish with a little over 1½ gallons (9.5 l.) specialty grain liquor. Add the dried malt extract and bittering hops and bring to a full and vigorous boil.

The total boil time will be 60 minutes. When 10 minutes remain, add Irish moss. After a total wort boil of 60 minutes, turn off the heat, add aroma hops and let steep 2 to 5 minutes. Then separate or strain out and sparge hops, and direct the hot wort into a sanitized fermenter to which 2 gallons (7.6 l.) of cold water have been added. If necessary, add additional cold water to achieve a 5-gallon (19 l.) batch size. Chill the wort to 65 to 70 degrees F (18–21 C). Aerate the cooled wort well. Add an active yeast culture and ferment for 4 to 6 days in the primary. Then transfer into a secondary fermenter, chill to 55 to 60 degrees F (13–15.5 C) if possible and add aroma hops. Allow to age for two weeks or more.

When secondary aging is complete, prime with sugar, bottle or keg. Let condition at temperatures above 60 degrees F (15.5 C) until clear and carbonated.

Silver Cup Winner

Red Brick Ale
Atlanta Brewing Co.
Atlanta, Georgia, USA

Red-amber color with a neat white head. Aroma is punctuated with the character of caramel malt and minimal grainy-cornlike character. Very little if any hop aroma emerges. Roast malt dominates the first flavor impressions, followed by the balance of caramel malt. Roast-malt astringency and bitterness complement and casually balance yet do not overpower hop bitterness. Well-defined chewy, malty character enjoins, yet is not cloying. Aftertaste is mild, malty with soft hop bitterness. Hop flavor is absent, yet not missed. Medium, pleasant body harmonizes and complements the overall impression of this interestingly complex ale.

Estimated profile based on tasting
Color: 10–13 SRM (20–26 EBC)
Bittering Units: 26–30

Bronze Cup Winner

Granville Island Pale Ale
Granville Island Brewing Co.
Vancouver, British Columbia, Canada

It is interesting to note that the label promotes this beer as a pale ale, yet it certainly is a great example of an English-style Brown Ale. Rich amber in color with orange hues. A rather neutral aroma with hints of maltiness. Medium-bodied with a light dry, clean finish. Minimally suggestive fruity-ester aroma. Overall flavor impresses as appropriately malty with subdued roasted or aromatic malt character. A light touch of caramel character and hint of roast/darker malt emerge quietly both in aroma and flavor. It certainly does not have any of the rich hop character and complexities of fruitiness of a "classic" pale ale. A smooth amber ale that impresses as an English-style Brown Ale in richness of malt character and subdued hop character. Exceptionally clean.

Estimated profile based on tasting
Alcohol by volume: 5% as indicated on label
Dolor: 11–13 SRM (22–26 EBC)
Bittering Units: 20–24

CATEGORY 9:
STRONG ALE

A. SUBCATEGORY:
ENGLISH OLD/STRONG ALE

Amber to copper to medium in color, English/Old Strong Ales are medium- to full-bodied with a malty sweetness. Fruity ester flavor and aroma should contribute to the character of this ale. Bitterness should be evident and balanced with malt and/or caramel sweetness. Alcohol types can be varied and complex. Chill haze is acceptable at low temperatures.

Original Gravity (°Plato):   1.055–1.075 (14–19 °Plato)
Apparent Extract–
Final Gravity (°Plato):   1.008–1.020 (2–5 °Plato)
Alcohol by weight (volume):   4.8–6.4% (6.0–8.0%)
Bitterness (IBU):   30–40
Color SRM (EBC):   10–16 (20–35 EBC)

B. SUBCATEGORY: STRONG "SCOTCH" ALE

Scotch Ales are overwhelmingly malty and full-bodied. Perception of hop bitterness is very low. Hop flavor and aroma are very low or none. Color ranges from deep copper to brown. The clean alcohol flavor balances the rich and dominant sweet maltiness in flavor and aroma. A caramel character is often a part of the profile. Fruity esters are generally at medium aromatic and flavor levels. A peaty-smoky character may be evident at low levels. Low diacetyl levels are acceptable. Chill haze is allowable at cold temperatures.

Original Gravity (°Plato):   1.072–1.085 (18–21 °Plato)
Apparent Extract–
Final Gravity (°Plato):   1.016–1.028 (4–7 °Plato)

Alcohol by weight (volume):   5.2–6.7% (6.2–8%)

Bitterness (IBU):   25–35

Color SRM (EBC):   10–25 (20–100 EBC)

C. SUBCATEGORY: OTHER STRONG ALES

Any style of beer can be made stronger than the classic style guidelines. The goal should be to reach a balance between the style's character and the additional alcohol. See this guide for specifics on the style being made stronger and identify the style when entering (for example: double alt, triple bock or quadruple Pilsener).

Gold Cup Winner

Star Brew 1000 Wheat Wine
Marin Brewing Company
1809 Larkspur Landing Circle
Larkspur, California, USA 94939
Brewmasters: Brendan J. Moylan and Arne Johnson
Established 1989
Production: 2,700 bbl. (3,200 hl.)

Winning beer competitions did not begin for this brewpub with the World Beer Cup. Marin's beers have been winners since they were first entered in the Great American Beer Festival in 1989. Inspired by his homebrewing interests and background in retail beer sales, brewmaster Brendan Moylan founded Marin County's first brewpub with Craig Tasley in 1989. Situated on the north side of the San Francisco Bay, the brewpub has attracted a strong following of beer enthusiasts and travelers with its eclectic list of quality craft-brewed beers.

The Marin Brewing Company, with its 130-seat restaurant and outdoor beer garden, approaches the beer business with a "good basic food" and fun-atmosphere philosophy.

Creative formulation is one of the more admirable aspects of the Marin Brewing Company's beer list. Where does that come from? One might naturally guess: homebrewing. Brendan's and Arne's love and appreciation for the homebrew hobby have led to their

Arne Johnson, Co-Brewmaster

Brendan Moylan

successful careers in the brewpub business. The brewpub also serves as a meeting place for the local homebrew club. You'll find the Marin Mountain Hoppers meeting and tasting their homebrews at 8:00 P.M. on the last Thursday of every month. This northern stretch of the San Francisco Bay area never had it so good.

Other products offered include Mr. Tam Pale Ale, Albion Amber, Marin Weiss, Marin Hefe Weiss, Doppel Weizen, Hefe Doppel Weizen, Pt. Reyes Porter, Old Dipsea Barleywine, Raspberry Trail Ale, Blueberry Ale, Stinson Beach Peach, St. Brendan's Irish Red Ale, Miwok Weizen Bock, Smoked Harvest Ale, Hoppy Holidaze Ale, Bodega Bay Bitter and Harvest Ale, as well as their World Cup winner, San Quentin's Breakout Stout. Star Brew 1000 Wheat Wine, inspired by friend Phil Moeller of the Rubicon Brewing Company, is only brewed every four years and only during a leap year.

Appearance is simply an amber color with an orange hue. It has a slight chill haze, but that's to be expected with a wheat wine. Aroma is delectably playful with a sweet caramel-toffee malt character married with an abundance of American hop aroma and nostril-warming alcohol. Gratefully the hop character is evident but does not overshadow the lascivious soft caramel malt. The flavor impression wallops the senses—this is a sky-high, wonderful, full-flavored, full-bodied beer with a soothing finish. Sensually alcoholic, the flavor is balanced with light caramel and malt followed with a wave of hop bitterness, first impressing as an intrusion, then appreciated for the balance of strength it offers to this skillfully executed brew. As big as this beer is, it is a wonder that it is not sickly sweet in the least. Quite a nightcap— Good night, Irene! A clean fermentation with only a mild hint of fruity esters. The final impression is all malt, hops and alcohol. A classic American invention.

- **Recipe for 5 U.S. gallons (19 liters) Star Brew 1000 Wheat Wine**
  Targets:
  Original Gravity: 1.098 (23.5)
  Final Gravity: 1.026 (6.5)
  Alcohol by volume: 9%
  Color: 11 SRM (22 EBC)
  Bittering Units: 50

ALL-GRAIN RECIPE AND PROCEDURE

8	lbs. (3.6 kg.)	American 2-row Klages pale malt
12	lbs. (5.5 kg.)	American wheat malt
20	**lbs. (9.1 kg.)**	**Total grains**

12.5 HBU (354 MBU) American Chinook shops (pellets)—105 minutes (bittering)

6 HBU (170 MBU) American Chinook hops (pellets)—
30 minutes (flavor)

1½ oz. (42 g.) American Cascade hops (pellets)—
steep in finished boiled wort for 2 to 3 minutes (aroma)

¼ tsp. Irish moss

¾ c. corn sugar for priming in bottles. Use ⅓ cup corn sugar if priming a keg.

Wyeast 1968 London ESB Ale yeast

Special notes for all-grain recipe: Because of the density of the boiling wort, it is difficult to calculate actual hop utilization using brewing formulas. Typical calculations for the prescribed hop dosage would yield about 60 bitterness units; however, it is estimated that actual bitterness will be closer to 50. Mash yields have been reduced to 70 percent efficiency due to the limitations of typical homebrewing systems. Sparge water is less than normally calculated so that total boiling volumes can be reasonably handled. Extended boiling time will increase carmelization of the wort, thus darkening the color from a calculated 8 SRM to a more likely 11 SRM.

A single-step infusion mash is employed to mash the grains. Add 20 quarts (19 l.) of 169-degree F (76 C) water to the crushed grain, stir, stabilize and hold the temperature at 152 degrees F (67 C) for 60 minutes.

After conversion, raise temperature to 167 degrees F (75 C), lauter and sparge with 5 gallons (19 l.) of 170-degree F (77 C) water. Collect about 8 gallons (30.5 l.) of runoff, add bittering hops and bring to a full and vigorous boil.

The total boil time will be 105 minutes. When 30 minutes remain, add flavor hops. When 10 minutes remain, add Irish moss. After a total wort boil of 105 minutes (or reducing the wort volume to just over 5 gallons), turn off the heat, add aroma hops and let steep 2 to 5 minutes. Then separate or strain out and sparge hops. Chill the wort to 65 to 70 degrees F (18–21 C) and direct into a sanitized fermenter. Aerate the cooled wort extremely well. Add a healthy amount of active yeast culture and ferment for 4 to 8 days in the primary. Then transfer into a secondary fermenter and maintain temperature. Secondary aging should last at least two to three months.

When secondary aging is complete, prime with sugar, bottle or keg. Let condition at temperatures above 60 degrees F (15.5 C) until clear and carbonated. This beer will take on its winning character as it approaches one year in age.

MALT-EXTRACT RECIPE AND PROCEDURE FOR STAR BREW 1000 WHEAT WINE

13½ lbs.	(5.9 kg.)	wheat malt extract syrup (50% wheat, 50% barley)
12.5 HBU	(354 MBU)	American Chinook hops (pellets)— 90 minutes (bittering)
6 HBU	(170 MBU)	American Chinook hops (pellets)— 30 minutes (flavor)
1½ oz.	(42 g.)	American Cascade hops (pellets)— steep in finished boiled wort for 2 to 3 minutes (aroma)
¼ tsp.		Irish moss
¾ c.		corn sugar for priming in bottles. Use ⅓ cup corn sugar if priming a keg.

Wyeast 1968 London ESB Ale yeast

Special notes for all-malt extract recipe: Because of the density of the boiling wort, it is difficult to calculate actual hop utilization using brewing formulas. Typical calculations for the prescribed hop dosage would yield about 65 bitterness units; however, it is estimated that actual bitterness will be closer to 50. With such a concentrated and long boil time used to reduce the wort volume, there will be carmelization of the wort, thus lending a caramel flavor and darkening the color from a calculated 8 SRM to a more likely 11 SRM.

Add the malt extract and bittering hops to 4 gallons of water and bring to a full and vigorous boil. The total boil time will be 90 minutes. When 30 minutes remain, add flavor hops. When 10 minutes remain, add Irish moss. After a total wort boil of 90 minutes (or a reduction of the wort to about 2½ to 3 gallons), turn off the heat, separate or strain out and sparge hops, and direct the hot wort into a sanitized fermenter to which 2 gallons (7.6 l.) of very cold water have been added. If necessary, add additional cold water to achieve a 5-gallon (19 l.) batch size. Chill the wort to 65 to 70 degrees F (18-21 C). Aerate the cooled wort extremely well. Add a healthy amount of active yeast culture and ferment for 4 to 8 days in the primary. Then transfer into a secondary fermenter and maintain temperature. Secondary aging should last at least two to three months.

When secondary aging is complete, prime with sugar, bottle or keg. Let condition at temperatures above 60 degrees F (15.5 C) until clear and carbonated. This beer will take on its winning character as it approaches one year in age.

Silver Cup Winner

None

Bronze Cup Winner

None

CATEGORY 10:
BARLEY WINE–STYLE ALE

Barley Wines range from tawny copper to dark brown in color and have a full body with high residual malty sweetness. Complexity of alcohols and fruity-ester characters are often high and counterbalanced by the perception of low to assertive bitterness and extraordinary alcohol content. Hop aroma and flavor may be minimal to very high. Diacetyl should be very low. A caramel and vinous aroma and flavor are part of the character. Chill haze is allowable at cold temperatures.

Original Gravity (°Plato):   1.090–1.120 (22.5–30.0 °Plato)
Apparent Extract–
Final Gravity (°Plato):   1.024–1.032 (6–8 °Plato)
Alcohol by weight (volume):   6.7–9.6% (8.4–12%)
Bitterness (IBU):   50–100
Color SRM (EBC):   14–22 (35–90 EBC)

Gold Cup Winner

Ozone Ale
Hubcap Brewery/Brewing Co.
143 E. Meadow Dr.
Vail, Colorado, USA 81658
Head Brewer: J. R. Rulapaugh
Established 1991
Production: 1,800 bbl. (2,100 hl.)

From the mountains and ski slopes of Vail, Colorado, head brewer J. R. Rulapaugh brews seventeen styles of ales for the love of full-flavored beers. There have been a lot of former homebrewers who have influenced beer character at

Lance Lucey

J. R. *Rulapaugh*, *Head Brewer*

this small brewery and brewpub, so much so that it is often hard to make the distinction between your own best home-brews and the beers brewed in their 7-barrel brewhouse here at 9,000 feet.

The popularity of Hubcap beers at certain times of the season maxes out their brewing capacity. They serve six brands on tap with several rotating seasonal specialties to address any beer enthusiast's mood swings. A limited amount of Hubcap ales are bottled and are available in select locations in Colorado, but as always, it is worth a visit to the source for the finest.

A single-step infusion mash is the procedure of choice for all their ales. They use English pale, Canadian organic pale and a combination of English and Belgian specialty malts. Hops come from England, the Czech Republic and the United States.

Ales regularly on tap are White River Wheat, Camp Hale Golden Ale, Ace Amber Ale, Beaver Tail Brown Ale, Vail Pale Ale and Rainbow Trout Stout. Powder Pig Porter, Ozone (barley wine–style) Ale, Tripel Lobotomy (brewed with Belgian ale yeast), Vailfest and Bagpipe Scotch Ale are formidable specialties gracing the spare taps on a rotating basis. Over the years the Hubcap Brewery has won several medals at the Great American Beer Festival, including two Gold Medals in the India Pale Ale category (1994 and 1995) with their Vail Pale Ale and a Silver Medal each for Beaver Tail Brown Ale in the Brown Ale category (1991) and Rainbow Trout Stout in the Dry Stout category (1992).

CHARACTER DESCRIPTION OF GOLD CUP–WINNING OZONE ALE

Deep red garnet ale with hints of orange in the hue. Fresh zesty and alcoholic aroma greets the senses. Kent Goldings hop aroma is surely evident but refrains from screaming. It is buttressed with a soft, gently stated toffee and caramel aroma. First flavor impression brings to mind the sweet love of caramel (but not a buttery caramel), becoming entangled with my own sneeze and hop bitterness. Perhaps roasted or black malt suggests itself in a very small way. Reluctant bitterness only goes as far as compensating for Ozone Ale's sweetness. A coconutlike ester weaves its wonderful and exotic essence into the overall balance. Hardly any harsh or higher-distracting alcohols or astringency. And oh, yes, even at lower altitudes the warmth of alcohol has its effect, but it is finely balanced with a soft body and mouth feel, caramel and bitterness. Toffeelike caramel character follows in flavor, but, as in the aroma, is not overdone. This well-balanced, full-bodied barley wine ale finishes pleasingly clean. Aftertaste lingers with a comforting bitterness and a hint of toffee. A wonderfully sexy barley wine, a cause for and contributor to celebration.

- ### Recipe for 5 U.S. gallons (19 liters) Ozone Ale
  Targets:
  Original Gravity: 1.092 (22)
  Final Gravity: 1.028 (7)
  Alcohol by volume: 8.5%

Color: 20 SRM (40 EBC)
Bittering Units: 88

ALL-GRAIN RECIPE AND PROCEDURE

16½ lbs.	(7.5 kg.)	English 2-row pale malt
¼ lb.	(114 g.)	Belgian Cara Vienne malt
½ lb.	(0.23 kg.)	English Carastan malt
½ lb.	(0.23 kg.)	English crystal malt—60 Lovibond
¼ lb.	(114 g.)	English crystal malt—135/165 Lovibond
18 lbs.	**(8.2 kg.)**	**Total grains**

16	HBU	(454 MBU) American Columbus hops (pellets)—105 minutes (bittering)
10	HBU	(284 MBU) English Kent Goldings hops (pellets)—30 minutes (flavor)
1½ oz.		(42 g.) American Willamette hops (whole)—dry-hop for 2 weeks (aroma)

| ¼ tsp. | Irish moss |
| ¾ c. | corn sugar for priming in bottles. Use ⅓ cup corn sugar if priming a keg. |

Wyeast 1187 Ringwood Ale yeast or Wyeast 1742 Swedish Ale yeast

Special notes for all-grain recipe: Because of the density of the boiling wort, it is difficult to calculate actual hop utilization using brewing formulas. Typical calculations for the prescribed hop dosage would yield about 98 bitterness units; however, it is estimated that actual bitterness will be closer to 88. Mash yields have been reduced to 70 percent efficiency due to the limitations of typical homebrewing systems. Sparge water is less than normally calculated so that total boiling volumes can be reasonably handled. Extended boiling time will increase carmelization of the wort, thus darkening the color from a calculated 18 SRM to a more likely 20 SRM.

*   *   *

A single-step infusion mash is employed to mash the grains. Add 18 quarts (17 l.) of 169-degree F (76 C) water to the crushed grain, stir, stabilize and hold the temperature at 152 degrees F (67 C) for 60 minutes.

After conversion, raise temperature to 167 degrees F (75 C), lauter and sparge with 5 gallons (19 l.) of 170-degree F (77 C) water. Collect about 8 gallons (30.5 l.) of runoff, add bittering hops and bring to a full and vigorous boil.

The total boil time will be 105 minutes. When 30 minutes remain, add flavor hops. When 10 minutes remain, add Irish moss. After a total wort boil of 105 minutes (or reducing the wort volume to just over 5 gallons), turn off the heat, then separate or strain out and sparge hops. Chill the wort to 70 degrees F (21 C) and direct into a sanitized fermenter. Aerate the cooled wort extremely well. Add a healthy amount of active yeast culture and ferment for 4 to 8 days in the primary. Then transfer into a secondary fermenter and maintain temperature for an additional two weeks. Then add aromatic dry hops and age for two more weeks. Transfer into another aging vessel, straining out the dry hops. Age for at least two to three months for additional smoothness.

When secondary aging is complete, prime with sugar, bottle or keg. Let condition at temperatures above 60 degrees F (15.5 C) until clear and carbonated. This beer will take on its winning character as it approaches one year in age.

MALT-EXTRACT RECIPE AND PROCEDURE FOR OZONE ALE

9½	lbs.	(4.3 kg.)	English light dried malt extract
½	lb.	(0.23 kg.)	English Carastan malt
¾	lb.	(0.34 kg.)	English crystal malt—60 Lovibond
½	lb.	(0.23 kg.)	English crystal malt—135/165 Lovibond
1¾	**lbs.**	**(0.8 kg.)**	**Total grains**

20   HBU (567 MBU) American Columbus hops (pellets)—75 minutes (bittering)

13.5 HBU   English Kent Goldings hops (pellets)—30 minutes (flavor)

1½ oz.   (42 g.) American Willamette hops (whole)—dry-hop for 2 weeks (aroma)

¼ tsp.   Irish moss

¾ c.   corn sugar for priming in bottles. Use ⅓ cup corn sugar if priming a keg.

Wyeast 1187 Ringwood Ale yeast or Wyeast 1742 Swedish Ale yeast

Steep crushed specialty grains in 1½ gallons (5.7 l.) water at 150 degrees F (65.5 C) for 30 minutes. Strain and sparge with enough 170-degree F (76.5 C) water to finish with 3 gallons (11.4 l.) specialty grain liquor. Add the dried malt extract and bittering hops and bring to a full and vigorous boil.

The total boil time will be 75 minutes. When 30 minutes remain, add flavor hops. When 10 minutes remain, add Irish moss. After a total wort boil of 75 minutes, turn off the heat, separate or strain out and sparge hops, and direct the hot wort into a sanitized fermenter to which 2 gallons (7.6 l.) of very cold water have been added. If necessary, add additional cold water to achieve a 5-gallon (19 l.) batch size. Chill the wort to 70 degrees F (21 C). Aerate the cooled wort extremely well. Add a healthy amount of active yeast culture and ferment for 4 to 8 days in the primary. Then transfer into a secondary fermenter and maintain temperature for an additional two weeks. Then add aromatic dry hops and age for two more weeks. Transfer into another aging vessel, straining out the dry hops. Age for at least two to three months for additional smoothness.

When secondary aging is complete, prime with sugar, bottle or keg. Let condition at temperatures above 60 degrees F (15.5 C) until clear and carbonated. This beer will take on its winning character as it approaches one year in age.

Silver Cup Winner

Big 12 Barley Wine
Little Apple Brewing Co.
Manhattan, Kansas, USA

Deep amber with red and orange hues. Explosively elegant caramel-malt aroma with complex fruitiness, slight butterscotch, slight banana, toffee and perhaps some hops. Hops are difficult to commit to in aroma with such complexity. Rich, full-bodied texture precedes the warming sparkle of alcohol, expressing its strength with the first sip. Bitterness is high enough that claiming to perceive any subtleties in numbers is folly. This is a malt-accented barley wine with a wonderful balance of quick-punch bitterness. The alcohol unsuccessfully attempts to smother other complex sensations of malt and the exotic accents of fermentation. Full-bodied. Full-textured. Full strength. Well balanced. The barest whisper of smokiness, perhaps from a handful of peat-smoked malt.

Estimated profile based on tasting
Color: 20–22 SRM (40–44 EBC)
Bittering Units: 50+ maybe up to 65

Bronze Cup Winner

Old Blue Granite Barley Wine
Blind Pig Brewing Co.
Temecula, California, USA

Deep coppery reddish brown barley wine sitting idly, visually calm, but with one sip . . . you only begin this great experience. Rich, deep aromatic caramel note contrasted with high sweet notes, yet not robustly sweet. Hop aroma is evident but not the predominant theme. Fruitiness is expressed as a faint wild cherry or plum. No evidence of bananas or apples. Mouth feel is full-bodied with a pale malt and cara-mel-like sweetness followed by an effusive dose of hop bit-terness compensating for the meaty, beaty, big and bouncy malt and body. Alcohol warmth and flavor are clean. No higher-solventlike alcohols detected. Subtle roast-malt bit-terness combines with the hop complexity, full body and sweet malt tones, lending a finished impression unlike any one component. Aftertaste is robust, estery, alcoholically warming, but not sharp. After all of this, one is no doubt enjoying the indulgence, and dare I conclude with any sense of legitimacy at this point that there is a faint sense of toffee? But then at this point I may not be able to make sense. Try it yourself and then tuck yourself in.

Brewery formulation uses 6 percent of various crystal malts and Chinook, Cascade and Centennial hops for bitter-ness, flavor and dry aroma hopping.

Estimated profile based on tasting
Original Gravity: 1.098 (23.5) indicated by the brewery
Final Gravity: 1.024 (6) indicated by the brewery
Alcohol by volume: 9.3% indicated by the brewery
Color: 15–16 SRM (30–32 EBC)
Bittering Units: 56–61(100 indicated by the brewery)

CATEGORY 11: ROBUST PORTER

Robust porters are black in color and have a roast malt flavor, but no roasted barley flavor. These porters have a sharp bitterness of black malt without a highly burnt/char-coal flavor. Robust porters range from a medium to full body and have a malty sweetness. Fruity esters should be evident and in proportional balance with roast malt and hop bitterness character. Hop flavor and aroma may vary from being negligible to medium in character.

Original Gravity (° Plato):	1.045–1.060 (11–15 °Plato)
Apparent Extract– Final Gravity (°Plato):	1.008–1.016 (2–4 °Plato)
Alcohol by weight (volume):	4.0–5.2% (5.0–6.5%)
Bitterness (IBU):	25–40
Color SRM (EBC):	30+(60+EBC)

Gold Cup Winner

Saint Brigid's Porter
Great Divide Brewing Company
2201 Arapahoe St.
Denver, Colorado, USA 80205
Brewmaster: Brian Dunn
Established 1993
Production: 8,000 bbl. (9,400 hl.)

Founded by homebrewer and current brewmaster Brian Dunn, the Great Divide Brewing Company is one of several downtown Denver microbreweries currently thriving on the quality of the beers they are producing. Established in an

Tara Dunn, Co-Founder/VP and Brian Dunn, Founding Brewer

old dairy building at an original capacity of 900 barrels per year, the Great Divide Brewing Company has since expanded to a capacity of 15,000 barrels with over thirteen employees. Brian and his wife, Tara, have their own special equity in their brewery, having had the inaugural brew take temporary priority over their honeymoon after their marriage on May 28, 1994. The first batch of beer was brewed on May 30, 1994, and was well received at its debut at the 1994 Colorado Brewers Festival.

Currently available in eleven states, Great Divide beers include Whitewater Wheat Ale, Bee Sting Honey Ale, Wild Raspberry Ale and Arapahoe Amber, as well as the award-winning Saint Brigid's Porter. Seasonal brews are also looked forward to by local beer enthusiasts.

The label proclaims that "Irish lore has it that St. Brigid, the second most well known Saint after St. Patrick, was credited with the amazing miracle of transforming her bathwater into beer for thirsty clerics. We at Great Divide feel that this deed is worthy of sainthood. Don't you?"

CHARACTER DESCRIPTION OF GOLD CUP–WINNING SAINT BRIGID'S PORTER

Deep dark brown color; actually quite opaque for all practical purposes with a rich, tan head. An exceptionally dry (not sweet) chocolatelike aroma with strong toasted-malt overtones. The strong cocoa character is this beer's memorable signature. A chocolate lover's beer dream come true. The first impression is of a medium-bodied beer with suggestions of dryness from roasted-malt astringency and fermentation. A grainy roast character follows and quickly diminishes as a clean semisweet chocolate flavor with caramel overtones emerges. Roast-grain bitterness is evident but quickly recedes. Fruitiness is not evident in flavor or aroma. No hop character to speak of, just a perfect balance of bitterness, creating a velvet sheen on the palate. Soft and mild, probably a midrange neutral American ale hop. Saint Brigid's Porter dances on the edge of the extreme,

feigning dark indulgence, but playing it safe, content to simply please. With its very clean fermented character, enjoy
the space you occupy and the beer you're with—that is the
primary message.

- **Recipe for 5 U.S. gallons (19 liters) Saint Brigid's
  Porter**
  Targets:
  Original Gravity: 1.060 (15)
  Final Gravity: 1.014 (3.5)
  Alcohol by volume: 6%
  Color: 75 or higher SRM (150 EBC)
  Bittering Units: 25

ALL-GRAIN RECIPE AND PROCEDURE

7	lbs. (3.2 kg.)	English 2-row pale malt
2	lbs. (0.9 kg.)	American Munich malt
1¼	lbs. (0.57 kg.)	American caramel malt—130 (or higher) Lovibond
½	lb. (0.23 kg.)	American wheat malt
½	lb. (0.23 kg.)	American Victory or other aromatic malt
1	lb. (0.45 kg.)	American chocolate malt
¼	lb. (114 g.)	American black malt
12½ lbs. (5.7 kg.)		**Total grains**

5.5 HBU (156 MBU) American Willamette hops (whole)—
           90 minutes (bittering)
2.5 HBU (71 MBU) American Tettnanger hops (whole)—
           45 minutes (flavor)

¼   tsp.   Irish moss
¾   c.   corn sugar for priming in bottles. Use ⅓ cup
           corn sugar if priming a keg.
Wyeast 1056 American Ale yeast or other neutral, low-
ester, high-attenuating yeast

A single-step infusion mash is employed to mash the grains. Add 12.5 quarts (11.9 l.) of 167-degree F (75 C) water to the crushed grain, stir, stabilize and hold the temperature at 150 degrees F (65.5 C) for 60 minutes.

After conversion, raise temperature to 167 degrees F (75 C), lauter and sparge with 4 gallons (15 l.) of 170-degree F (77 C) water. Collect about 6 gallons (22.8 l.) of runoff, add bittering hops and bring to a full and vigorous boil.

The total boil time will be 90 minutes. When 45 minutes remain, add flavor hops. When 10 minutes remain, add Irish moss. After a total wort boil of 90 minutes (reducing the wort volume to just over 5 gallons), turn off the heat, then separate or strain out and sparge hops. Chill the wort to 65 to 70 degrees F (18–21 C) and direct into a sanitized fermenter. Aerate the cooled wort well. Add an active yeast culture and ferment for 4 to 6 days at temperatures between 60 and 65 degrees F (15.5–18.5 C) in the primary. Then transfer into a secondary fermenter and chill to 60 degrees F (13–15.5 C) if possible.

When secondary aging is complete, prime with sugar, bottle or keg. Let condition at temperatures above 60 degrees F (15.5 C) until clear and carbonated.

MASH-EXTRACT RECIPE AND PROCEDURE FOR SAINT BRIGID'S PORTER

4	lbs.	(1.8 kg.)	English dried amber malt extract
1½	lbs.	(0.7 kg.)	English 2-row pale malt
1	lb.	(0.45 kg.)	American Munich malt
1¼	lbs.	(0.57 kg.)	American caramel malt—130 (or higher) Lovibond
½	lb.	(0.23 kg.)	American wheat malt
½	lb.	(0.23 kg.)	American Victory or other aromatic malt
1	lb.	(0.45 kg.)	American chocolate malt
¼	lb.	(114 g.)	American black malt
6	**lbs.**	**(2.7 kg.)**	**Total grains**

6.25 HBU (177 MBU) American Willamette hops
            (whole)—75 minutes (bittering)
2.5   HBU (71 MBU) American Tettnanger hops (whole)—
            45 minutes (flavor)

¼ tsp.   Irish moss
¾ c.   corn sugar for priming in bottles. Use ⅓ cup
            corn sugar if priming a keg.
Wyeast 1056 American Ale yeast or other neutral, low-
ester, high-attenuating yeast

A single-step infusion mash is employed to mash the
grains. Add 6 quarts (5.7 l.) of 172-degree F (78 C) water to
the crushed grain, stir, stabilize and hold the temperature
at 156 degrees F (69 C) for 60 minutes.

After conversion, raise temperature to 167 degrees F (75
C), lauter and sparge with 2.5 gallons (2.4 l.) of 170-degree
F (77 C) water. Collect about 3 gallons (11.5 l.) of runoff.
Add malt extract and bittering hops and bring to a full and
vigorous boil.

The total boil time will be 75 minutes. When 45 minutes
remain, add flavor hops. When 10 minutes remain, add Irish
moss. After a total wort boil of 75 minutes (reducing the
wort volume to just over 5 gallons), turn off the heat, then
separate or strain out and sparge hops, and direct the hot
wort into a sanitized fermenter to which 2 gallons (7.6 l.) of
cold water have been added. If necessary, add additional
cold water to achieve a 5-gallon (19 l.) batch size. Chill the
wort to 70 degrees F (21 C). Aerate the cooled wort well.
Add an active yeast culture and ferment for 4 to 6 days at
temperatures between 60 and 65 degrees F (15.5–18.5 C) in
the primary. Then transfer into a secondary fermenter, chill
to 60 degrees F (13–15.5 C) if possible.

When secondary aging is complete, prime with sugar, bottle or keg. Let condition at temperatures above 60 degrees
F (15.5 C) until clear and carbonated.

Silver Cup Winner

BridgePort Porter
BridgePort Brewing Co.
Portland, Oregon, USA

Deep, dark and opaque bottle-conditioned brew. Wonderful roasted-malt and cocoa aroma that is rich without being excessive. Pleasant fruitiness adds to the exotic complexity of this enticingly aromatic porter; has a special combination of characters found only in bottle-conditioned ales (and lagers). Medium-bodied mouth feel with an extraordinarily clean finish. Malt and hops linger joyously in the flavor along with a jig dance of fruitiness from the bottle-conditioned yeast character. Aftertaste attempts to be light-hearted but yields to the complexity of roast malt, bottle conditioning and well-balanced dose of hops. Hop effect is medium; neither soft nor harsh and American in nature. Richly satisfying porter. Extraordinarily clean.

Estimated profile based on tasting
Color: 40+ SRM (80+ EBC)
Bittering Units: 35–38

Bronze Cup Winner

The Edmund Fitzgerald Porter
Great Lakes Brewing Co.
Cleveland, Ohio, USA

Dark brown and simply opaque. Roast, chocolate and toasted malt characters blossom in the aroma. Flavor has a pleasant degree of black-roast malt astringency which lends some bitterness and dryness, adding to a reassuring porter experience. The dry, controlled bitterness is only a secondary theme. Though the nature of the bitterness is almost harsh, it seems to be very appropriate given the complexity of sweet malt (caramel), chocolate malt and thematic alelike fruitiness. Chocolate malt is not sweet chocolate but rather a balanced and suggestive bitter chocolate.

Overall taste impression is fresh, clean, assertive and mysteriously alluring. An excellent medium-bodied, full-flavored porter with a dry finish.

Estimated profile based on tasting
Color: 28+ SRM (56+ EBC)
Bittering Units: 38–43

CATEGORY 12: BROWN PORTER

Brown Porters are medium to dark brown in color and may have a red tint. No roasted-barley or strong burnt-malt character should be perceived. Low to medium malt sweetness is acceptable along with medium hop bitterness. This is a light- to medium-bodied beer. Fruity esters are acceptable. Hop flavor and aroma may vary from being negligible to medium in character.

Original Gravity (°Plato):   1.045–1.060 (11–15 °Plato)
Apparent Extract–
Final Gravity (°Plato):   1.008–1.016 (2–4 °Plato)
Alcohol by weight (volume):   4.0–5.2% (5.0–6.5%)
Bitterness (IBU): 20–30
Color SRM (EBC):   20–35+ (40–70+ EBC)

Stephanie A. Weins

(*l. to r.*) Kent Taylor, Owner; Stephanie Weins, Owner and G.M.; and Dave Miller, Owner and Brewmaster

Gold Cup Winner

St. Charles Porter
Blackstone Restaurant and Brewery
1918 West End Ave.
Nashville, Tennessee, USA 37203
Brewmaster: David Miller
Established 1994
Production: 1,000 bbl. (1,200 hl.)

In the heart of Nashville, some might say there is an inspiration for the southeastern brewpub and restaurant scene. Founded in 1994 by Kent Taylor and Stephanie Weins, the Blackstone Restaurant and Brewery began as a homebrewer's dream, although one not without several obstacles that had to be overcome in making the dream a reality: the need to raise money and secure a location and, for Kent, to work on the real estate closing twenty-four hours after surgery for a herniated disk. But the founders prevailed, and the Blackstone fortunes soon found its brewmaster, award-winning homebrew pioneer and expert Dave Miller. Dave was the American Homebrewers Association's 1981 Homebrewer of the Year and is the author of several books, including *The Homebrewing Guide* (Storey Communications), *Complete Handbook of Homebrewing* (Garden Way) and *Pilsener* (Brewers Publications' Classic Beer Style Series). Miller, formerly the original brewmaster of The St. Louis Brewery in St. Louis, Missouri, moved to Nashville to begin brewing his award-winning beer there.

Blackstone's beers attempt to present a variety of beer styles from around the world, using hops and malts from several different countries. The brewery claims its own unique interpretation of these classic styles.

St. Charles Porter is based on Kent Taylor's prizewinning homebrew recipe and Dave Miller's adaptation of it. St. Charles Porter was born from these two different recipes whose batches were combined. It was so good, Miller simply summed the two recipes and divided by two.

Brewmaster's specials have included Nut Brown Ale, Spiced Christmas Ale, Oatmeal Stout, Irish Stout, German Hefeweizen, Belgian Wit, Raspberry Wheat, Scotch Ale, Oktoberfest, California Common, Amber, India Pale Ale, Extra Special Bitter, Altbier, American wheat beers, English mild and Pilsener.

CHARACTER DESCRIPTION OF GOLD CUP–WINNING ST. CHARLES PORTER

Brown color with a red hue. Very sweet malty aroma with accent of toasted malt and chocolate character. Along with low-profile fruitiness, the aroma promises a well-balanced flavor. Hops not notable in aroma. The initial flavor impression suggests roasted-malt astringency and bitterness but not to an overwhelming degree that would distract from the finish. Small amount of black-malt character accents flavor impression. American-type hop flavor is evident to the perceptive palate, barely suggested and delicately accomplished. The impression of hop bitterness is low, likely due to the medium to full body.

- **Recipe for 5 U.S. gallons (19 liters) St. Charles Porter**
  Targets:
  Original Gravity: 1.052 (13)
  Final Gravity: 1.014 (3.5)
  Alcohol by volume: 5.3%
  Color: 25 SRM (50 EBC)
  Bittering Units: 30

ALL-GRAIN RECIPE AND PROCEDURE

7½ lbs.	(3.4 kg.)	American 2-row pale malt
¾ lb.	(0.34 kg.)	flaked barley
¾ lb.	(0.34 kg.)	English crystal malt—60 Lovibond
½ lb.	(0.23 kg.)	Belgian Special "B" malt
½ lb.	(0.23 kg.)	American Victory or other (Biscuit, etc.) aromatic malt
½ lb.	(0.23 kg.)	English chocolate malt
¼ lb.	(114 g.)	English black malt
10¾ lbs. (4.9 kg.)		**Total grains**

12.5 HBU (354 MBU) American Willamette hops (pellets)—30 minutes (bittering)

2.5 HBU (71 MBU) American Centennial hops (pellets)—10 minutes (flavor)

¼ tsp. Irish moss

¾ c. corn sugar for priming in bottles. Use ⅓ cup corn sugar if priming a keg.

Wyeast 1056 American Ale yeast

A single-step infusion mash is employed to mash the grains. Add 11 quarts (10.5 l.) of 171-degree F (77 C) water to the crushed grain, stir, stabilize and hold the temperature at 154 degrees F (68 C) for 60 minutes.

After conversion, raise temperature to 167 degrees F (75 C), lauter and sparge with 4 gallons (15 l.) of 170-degree F (77 C) water. Collect about 5.5 gallons (21 l.) of runoff and bring to a full and vigorous boil. Do not add bittering hops yet.

The total boil time will be 75 minutes. When 30 minutes remain, add bittering hops. When 10 minutes remain, add flavor hops and Irish moss. After a total wort boil of 75 minutes (reducing the wort volume to just over 5 gallons), turn off the heat, then separate or strain out and sparge hops. Chill the wort to 65 to 70 degrees F (18–21 C) and direct into a sanitized fermenter. Aerate the cooled wort well. Add an active yeast culture and ferment for 4 to 6 days in the primary. Then transfer into a secondary fermenter, chill to 55 to 60 degrees F (13–15.5 C) and let secondary finish fermentation and cold aging for another two weeks.

When secondary aging is complete, prime with sugar, bottle or keg. Let condition at temperatures above 60 degrees F (15.5 C) until clear and carbonated.

MASH-EXTRACT RECIPE AND PROCEDURE FOR ST. CHARLES PORTER

3½ lbs.	(1.6 kg.)	English light dried malt extract
1¾ lbs.	(0.80 kg.)	American 2-row pale malt
¾ lb.	(0.34 kg.)	flaked barley
¾ lb.	(0.34 kg.)	English crystal malt—60 Lovibond
½ lb.	(0.23 kg.)	Belgian Special "B" malt
½ lb.	(0.23 kg.)	American Victory or other (Biscuit, etc.) aromatic malt
½ lb.	(0.23 kg.)	English chocolate malt
¼ lb.	(114 g.)	English black malt
5 lbs.	**(2.3 kg.)**	**Total grains**

15   HBU (425 MBU) American Willamette hops (pellets)—30 minutes (bittering)

2.5 HBU (71 MBU) American Centennial hops (pellets)—10 minutes (flavor)

¼ tsp. Irish moss

¾ c. corn sugar for priming in bottles. Use ⅓ cup corn sugar if priming a keg.

Wyeast 1056 American Ale yeast

A single-step infusion mash is employed to mash the grains. Add 5 quarts (4.75 l.) of 171-degree F (77 C) water to the crushed grain, stir, stabilize and hold the temperature at 154 degrees F (68 C) for 60 minutes.

After conversion, raise temperature to 167 degrees F (75 C), lauter and sparge with 2 gallons (7.6 l.) of 170-degree F (77 C) water. Collect about 2.5 gallons (9.5 l.) of runoff. Add malt extract and bring to a full and vigorous boil. Do not add bittering hops. The total boil time will be 60 minutes. When 30 minutes remain, add bittering hops. When 10 minutes remain, add flavor hops and Irish moss. After a total wort boil of 60 minutes (reducing the wort volume to just over 5 gallons), turn off the heat, then separate or strain out and sparge hops. Chill the wort to 65 to 70 degrees F (18–21 C) and direct into a sanitized fermenter. Aerate the cooled wort well. Add an active yeast culture and ferment for 4 to 6 days in the primary. Then transfer into a secondary fermenter, chill to 55 to 60 degrees F (13–15.5 C) and let secondary finish fermentation and cold aging for another two weeks.

When secondary aging is complete, prime with sugar, bottle or keg. Let condition at temperatures above 60 degrees F (15.5 C) until clear and carbonated.

Silver Cup Winner

Haystack Black
Portland Brewing Co.
Portland, Oregon, USA

Rich, dark cocoa brown with overall red hue. Reassuring alelike fruitiness blends with a refreshing floral hop aroma. Though moderate in intensity, the relative assertiveness of the roast malt and black malt is a signature of the overall flavor and aroma characters. Dark caramel malt also contributes to flavor and aroma. The roasted and black malts also add bitterness and astringency, but impressions quickly move on to caramel sweetness, then to the tingling of alcohol. Fruity character comes through in the flavor while hop bitterness is soft and not too assertive. A gentle balance of roast malt, hop bitterness, fruity alelike character and malt sweetness combines with a medium-bodied mouth feel and a medium-dry finish. A porter by anyone's standard.

Brewery formulation uses light and dark caramel malts with Nugget, Willamette and Kent Goldings hops used for bitterness, flavor and aroma.

Estimated profile based on tasting
Original Gravity: 1.052 (13.0) indicated by the brewery
Final Gravity: 1.016 (4.0)
Alcohol by volume: 4.8%
Color: 20–24 SRM (40–48 EBC)
Bittering Units: 27–31 (27 indicated by the brewery)

Bronze Cup Winner

Stoddard's Porter
Stoddard's Brewhouse and Eatery
Sunnyvale, California, USA

Deep dark opaque ale with a hint of red. Cocoa and roasted-malt character in aroma with a butterscotch complement. A brisk hop aroma emerges, adding to an overall enchanting complexity. Medium-bodied mouth feel is followed with a remarkable dry finish due to the astringency the roast malt contributes. This enhances the drinkability, especially for those who enjoy medium- or lighter-bodied porters. Flavor is also reminiscent of a hazel-nutty, almost biscuitlike malt character. Not a particularly hoppy porter. Aftertaste is sweet and clean with a refreshing clean palate of subtle roast-malt bitterness. Pleasant and meaningful drinkability.

Brewery formulation uses English crystal, brown and chocolate malts with English Fuggles and German Hallertauer hops for bitterness, flavor and aroma.

Estimated profile based on tasting
Original Gravity: 1.052 (13)
Final Gravity: 1.016 (4.1)
Alcohol by volume: 4.5%
Color: 50+ SRM (100+ EBC) (about 40 SRM indicated by the brewery)
Bittering Units: 30–32 (25 indicated by the brewery)

CATEGORY 13: SWEET STOUT

Sweet Stouts, or Cream Stouts, have less roasted bitter flavor and more full-bodied mouth feel than Dry Stouts. The style can be given more body with milk sugar (lactose) before bottling. Malt sweetness, chocolate and caramel flavor should dominate the flavor profile. Hops should balance sweetness without contributing apparent flavor or aroma.

Original Gravity (°Plato):	1.045–1.056 (11–14 °Plato)
Apparent Extract– Final Gravity (°Plato):	1.012–1.020 (3–5 °Plato)
Alcohol by weight (volume):	2.5–5% (3–6%)
Bitterness (IBU):	15–25
Color SRM (EBC):	40+ (150+ EBC)

Gold Cup Winner

Mackeson XXX Stout
Whitbread Beer Company
Porter Tun House
500 Capability Green Luton
Bedfordshire, LU1 3LS, England
Brewmaster: Michael Howard
Original Mackeson Brewery established 1669
Production: 4.3 million bbl. (5 million hl.)

"It looks good, it tastes good . . . and by golly, it does you good!" With this promotion and others during the post–World War II era, Mackeson Stout sustained its legendary status as the one and original sweet "milk" stout.

The original Mackeson brewery dates back to 1669, when records show that a James Pashley mortgaged his brew-

Mike Howard, Brewmaster

Paul Abrams

house to another merchant for 50 pounds sterling. The brewery passed from one owner to another until 1790 when Henry Mackeson, then a wine merchant, bought the brewery. The brewery prospered during and after the Napoleonic Wars, brewing a range of cask beers such as Bitter, Hythe Guinea Ale, Light Pale Ale, Pale Ale, Porter and "Specialite" Anglo-Lager beer. But it wasn't until 1907, when the brewery launched a brand-new and innovative product called Mackeson's Milk Stout, that success and a place in brewing history was assured for the name Mackeson.

Individuals in England first applied for patents for a beer based on milk and malt in 1875. By the late 1890s many ideas had been tested on the practicality of using peptonized separated milk in lieu of water for the mashing process. But it wasn't until 1907 when the small brewery in Hythe, County Kent, actually created a product using milk sugars as an ingredient. Thus was born and patented the term "milk stout," an exclusive product of the Mackeson & Co. Brewery. The popularity of the product led to infringements on their patent and trademarks and the invention of such terms as "cream stout" to circumvent trademarks and patents.

Whitbread acquired the Mackeson Brewery in 1929, then brewing a full variety of cask and bottled beers such as Pale Ale, Light Pale Ale, India Pale Ale, SB Pale Ale, Old Ale, Dark Ale, Nut Brown Ale, Strong Ale, Old Kentish XXXXX Strong Ale, Stout and Milk Stout. Whitbread thought of discontinuing Mackeson Milk Stout, but while it considered the idea, the product proved wildly popular in test markets. Its continuation was assured.

World War II approached, exploded and ended. The production of Mackeson Mild Stout was never disrupted. Postwar attitudes toward the exaggerated benefits advertised by "milk stout" brewers helped bring an end to the use of the word "milk" with stout. Despite missing the word "milk" on the label, Mackeson Stout became one of the biggest successes for Whitbread, appreciated by those with a preference for sweeter beer. In 1961 sales of Mackeson Stout rose to nearly 500,000 U.S. barrels (585,000 hl.).

The original Hythe Brewery in Kent was abandoned in 1968. Operations for stout production were moved to Whitbread's Samlesbury brewery. In the heyday of the 1960s Mackeson Stout was exported to over sixty countries and brewed under license in Belgium, Saint Lucia, Jamaica, New Zealand and Singapore. Though not nearly as popular now, Mackeson Stout continues to be brewed with a small proportion of lactose sugar for added body and sweetness. It remains the leader and benchmark for all beer aficionados who appreciate the one and original "milk" stout.

CHARACTER DESCRIPTION OF GOLD CUP-WINNING MACKESON XXX STOUT

Rich black color layered with an equally rich brown head. Sweet malt aroma accompanied by a roasted chocolate-malt character presents a first impression. There is very little fruitiness. The overall character is clean and simply stated with no evidence of hop aroma. Flavor is initiated with a suggestion of black roasted, smoky malts. But the predominant impression is a complex malty sweetness accompanied by a light underlying bitterness. Mouth feel is full-bodied. The progression is from sweet to chocolate to biscuit-toasted malt characters, followed with a finishing touch of roasted-burnt malt and hop bitterness.

Mackeson's "secret" is the addition of both lactose and sucrose for body and sweetness. The beer is then pasteurized to prevent further fermentation, a step most homebrewers would rather avoid. Crystal and aromatic malts can be substituted, but the end result will never be quite the same. Heat pasteurization would be the only true means of coming close to authentically reproducing much of the character of this fabled and classic sweet stout.

- **Recipe for 5 U.S. gallons (19 liters) Mackeson XXX Stout**

Targets:
Original Gravity: 1.058 (14.5)
Final Gravity: 1.022 (5.5)
Alcohol by volume: 4.6%
Color: 135 SRM (270 EBC)
Bittering Units: 25

ALL-GRAIN AND ADJUNCT RECIPE AND PROCEDURE

7	lbs.	(3.2 kg.)	English 2-row pale malt
½	lb.	(0.23 kg.)	American Victory or other aromatic malt
½	lb.	(0.23 kg.)	English chocolate malt
1¾	lbs.	(0.8 kg.)	high-maltose corn syrup
¼	lb.	(114 g.)	highly caramelized corn syrup
8	**lbs.**	**(3.2 kg.)**	**Total grains**

6.5	HBU (184 MBU)	English Kent Goldings hops (pellets)—60 minutes (bittering)

¼	tsp.		Irish moss
4	oz.	(114 g.)	sucrose
2	oz.	(57 g.)	lactose

Wyeast 1338 European Ale yeast or 1968 London ESB can be recommended; low esters, low attenuation, emphasizes maltiness

Special equipment: CO_2 tank and regulator, 5-gallon keg, counterpressure bottling system.

A single-step infusion mash is employed to mash the grains. Add 8 quarts (7.6 l.) of 175-degree F (80 C) water to the crushed grain, stir, stabilize and hold the temperature at 158 degrees F (70 C) for 60 minutes.

After conversion, raise temperature to 167 degrees F (75 C), lauter and sparge with 4.5 gallons (17 l.) of 170-degree

F (77 C) water. Collect about 5.5 gallons (21 l.) of runoff, add corn syrups and bittering hops, and bring to a full and vigorous boil.

The total boil time will be 60 minutes. When 10 minutes remain, add Irish moss. After a total wort boil of 60 minutes (reducing the wort volume to just over 5 gallons), turn off the heat, then separate or strain out and sparge hops. Chill the wort to 65 to 70 degrees F (18–21 C) and direct into a sanitized fermenter. Aerate the cooled wort well. Add an active yeast culture and ferment for 4 to 6 days in the primary. Then transfer into a secondary fermenter and age for three weeks.

When secondary aging is complete, chill the beer to 33 degrees F (0 C). Dissolve the sucrose and lactose in a small amount of boiling water and add it to the beer, then transfer the beer into a 5-gallon keg. Artificially carbonate immediately by applying CO_2 pressure and dissolving the gas into solution. (See pages 183–87, The Home Brewer's Companion.) Using counterpressure bottle equipment, fill bottles. Fill to within one half to three quarters of an inch of the bottle top. At this point it is extremely important to thoroughly pasteurize each bottle by raising the temperature of the beer to 151 degrees F (66 C) for a period of 17 minutes. This is equivalent to about 120 pasteurization units, recommended for pasteurization of nonalcoholic beverages with fermentable carbohydrates present. Warning: If beer is not thoroughly pasteurized, yeast will ferment the remaining sucrose and cause excessive pressure, resulting in exploding bottles.

ALL-GRAIN RECIPE AND PROCEDURE FOR MACKESON XXX STOUT WITHOUT PASTEURIZATION

6¾	lbs.	(3.1 kg.)	English 2-row pale malt
½	lb.	(0.23 kg.)	American Victory or other aromatic malt
½	lb.	(0.23 kg.)	English chocolate malt
½	lb.	(0.23 kg.)	English black roast malt

1	lb.	(0.45 kg.)	English crystal malt—40 Lovibond
1	lb.	(0.45 kg.)	English crystal malt—135/165 Lovibond
1	lb.	(0.45 kg.)	American Cara-Pils malt
11¼ lbs.		**(5.1 kg.)**	**Total grains**

6.5 HBU (184 MBU) English Kent Goldings hops (pellets)—75 minutes (bittering)

¼ tsp. Irish moss
¾ c. corn sugar for priming in bottles. Use ⅓ cup corn sugar if priming a keg.

Wyeast 1338 European Ale yeast or 1968 London ESB can be recommended; low esters, low attenuation, emphasizes maltiness

A single-step infusion mash is employed to mash the grains. Add 11 quarts (10.5 l.) of 175-degree F (80 C) water to the crushed grain, stir, stabilize and hold the temperature at 158 degrees F (70 C) for 60 minutes.

After conversion, raise temperature to 167 degrees F (75 C), lauter and sparge with 4.5 gallons (17 l.) of 170-degree F (77 C) water. Collect 6 gallons (21 l.) of runoff, add corn syrups and bittering hops, and bring to a full and vigorous boil.

The total boil time will be 75 minutes. When 10 minutes remain add Irish moss. After a total wort boil of 75 minutes (reducing the wort volume to just over 5 gallons), turn off the heat, then separate or strain out and sparge hops. Chill the wort to 65 to 70 degrees F (18–21 C) and direct into a sanitized fermenter. Aerate the cooled wort well. Add an active yeast culture and ferment for 4 to 6 days in the primary. Then transfer into a secondary fermenter and age for three weeks.

When secondary aging is complete, prime with sugar, bottle or keg. Let condition at temperatures above 60 degrees F (15.5 C) until clear and carbonated.

MALT-EXTRACT RECIPE AND PROCEDURE

5	lbs.	(2.3 kg.)	English dark dried malt extract
½	lb.	(0.23 kg.)	English chocolate malt
½	lb.	(0.23 kg.)	English black roast malt
1	lb.	(0.45 kg.)	English crystal malt—40 Lovibond
1	lb.	(0.45 kg.)	English crystal malt—135/165 Lovibond
3	**lbs. (1.4 kg.)**	**Total grains**

8.5   HBU   (241 MBU) English Kent Goldings hops (pellets)—60 minutes (bittering)

¼   tsp.   Irish moss
¾   c.   corn sugar for priming in bottles. Use ⅓ cup corn sugar if priming a keg.

Wyeast 1338 European Ale yeast or 1968 London ESB can be recommended; low esters, low attenuation, emphasizes maltiness

Steep crushed speciality grains in 1½ gallons (5.7 l.) water at 150 degrees F (65.5 C) for 30 minutes. Strain and sparge with enough 170-degree F (76.5 C) water to finish with a little over 2½ gallons (9.5 l.) specialty grain liquor. Add the dried malt extract and bittering hops and bring to a full and vigorous boil.

The total boil time will be 60 minutes. When 10 minutes remain, add Irish moss. After a total wort boil of 60 minutes, turn off the heat, separate or strain out and sparge hops, and direct the hot wort into a sanitized fermenter to which 2 gallons (7.6 l.) of cold water have been added. If necessary, add additional cold water to achieve a 5-gallon (19–l.) batch size. Chill the wort to 70 degrees F (21 C). Aerate the cooled wort well. Add an active yeast culture and ferment for 4 to 6 days in the primary. Then transfer into a secondary and allow to age for two weeks or more.

When secondary aging is complete, prime with sugar, bottle or keg. Let condition at temperatures above 60 degrees F (15.5 C) until clear and carbonated.

Silver Cup Winner

Stillwater Cream Stout
Colorado Brewing Co.
Thornton, Colorado, USA

Deep, dark brown opaque ale. Sweet cocoa aroma emerges from combination of caramel and chocolate malts. Weaving itself into this tapestry stout, roasted barley and/or malt lends a pleasant and slightly smoky character to the aroma. The earthy foundation of roasted malts is complemented by a perfect balance of sweet caramel malts. Complexity continues with the suggestion of hazelnut/biscuitlike character. First flavor impression is a wonderfully clean character, yet full of roasted- and pale-malt complexity, balanced with nonintrusive hop character. This is a malt-based stout that is gentle and not satiatingly sweet. An extremely drinkable stout. Medium-bodied mouth feel is in harmony with an austere complexity worthy of World Beer Cup recognition. Aftertaste is not strong or bitter; rather it is like a gentle, well-made cup of smooth, satisfying coffee. An excellent example of the style, though not as sweet as the Gold Cup winner.

Brewery formulation uses crystal, chocolate, black and Cara-Pils malts and roast barley along with English Fuggles and East Kent Goldings hops for bitterness, flavor and aroma.

Estimated profile based on tasting
Original Gravity: 1.056 (14) indicated by the brewery
Final Gravity: 1.018 (4.5) indicated by the brewery
Alcohol by volume: 4.3% indicated by the brewery
Color: 30+ SRM (60+ EBC) (100 SRM indicated by the brewery)
Bittering Units: 27–28 (24 indicated by the brewery)

Bronze Cup Winner

None

CATEGORY 14: OATMEAL STOUT

Oatmeal Stouts typically include oatmeal in their grist, which results in a pleasant, full flavor and smooth profile that is rich without being grainy. Roasted malt character of caramel and chocolate should be evident, smooth and not bitter. Bitterness is moderate, not high. Hop flavor and aroma are optional but should not overpower the overall balance. This is a medium- to full-bodied, minimally fruity-estery beer.

Original Gravity (°Plato):	1.038–1.056 (9.5–14 °Plato)
Apparent Extract–Final Gravity (°Plato):	1.008–1.020 (2–5 °Plato)
Alcohol by weight (volume):	3.0–4.8% (3.8–6%)
Bitterness (IBU):	20–40
Color SRM (EBC):	20+ (80+ EBC)

Gold Cup Winner

Zoser Oatmeal Stout
Oasis Brewery and Restaurant
1095 Canyon Blvd.
Boulder, Colorado, USA 80302
Brewmaster: Bill Sherwood
Established 1992
Production: 9,000 bbl. (10,500 hl.) combined total of brew-pub and brewery annex

It's the brewpub that's four blocks from the office; my office, that is. Kind of a homebrewer's dream, if you like homebrew and you're dreaming my dream. You'll always find an unusual mix of people at the Oasis: veteran and new homebrewers, beer enthusiasts, college students, pool guppies and people who simply enjoy being people. The Oasis Brewery is no new-comer to the medal hall of fame. Under the worthy reign of brewmaster Bill Sherwood and owners George and Lynne Hanna, it has been winning Gold, Silver and Bronze medals at the Great American Beer Festi-val since it opened. The brewpub's 7-barrel brew-house consistently churns out acclaimed Capstone ESB, Zoser Oatmeal Stout, Tut Brown Ale, Oasis Pale Ale and Scarab Red, as well as

George Hanna

William Sherwood, Brewmaster

seasonal specialities such as Blueberry Ale, Lemon Ale, Fest Beer, Bocks, Doppelbocks and many more full-flavored brews. With pool tables on the second floor and great dining on the first, Oasis has always been a meeting place for brew folks.

Homebrewers have even had the opportunity to brew their recipes at the Oasis. Call the Oasis a homebrew-friendly brewpub. The beers and the ambience speak for themselves.

CHARACTER DESCRIPTION OF GOLD CUP-WINNING ZOSER OATMEAL STOUT

Opaque, deep dark brown. There's a rich brown head and some foam clinging to the glass. There's no light passing through this stout. Very appealing and rich in appearance. Aroma suggests caramel, roast barley and chocolate malts, with a toffeelike character. A noncitrusy hop aroma suggests itself, but fades quickly as it blends with malt aromatics. The flavor comes on strong with a rich chocolatelike texture balanced with the appropriate medium to full body of a perfect oatmeal stout. Roasted-malt-and-barley character comes through with some contribution of acidity and dry, roast astringency. This is actually quite refreshing, balancing the full body and malty sweetness of Zoser. No diacetyl, some sweet fruitiness in aroma, but not banana. Though full-bodied, its finish and aftertaste are clean and dry. Sweetness does not linger, but the astringency of roast malts does.

- **Recipe for 5 U.S. gallons (19 liters) Zoser Oatmeal Stout**
  Targets:
  Original Gravity: 1.063 (13.5)
  Final Gravity: 1.016 (4)
  Alcohol by volume: 6.9%
  Color: Black SRM (Black EBC)
  Bittering Units: 29

ALL-GRAIN RECIPE AND PROCEDURE

10	lbs. (4.5 kg.)	American 2-row pale malt
½	lb. (0.23 kg.)	American roasted barley
1	lb. (0.45 kg.)	American caramel malt—130 Lovibond
¾	lb. (0.34 kg.)	oatmeal
¾	lb. (0.34 kg.)	American chocolate malt
¼	lb. (114 g.)	American black malt
13¼ lbs. (6.0 kg.)		**Total grains**

3 HBU (85 MBU) Cascade hops (pellets)—90 minutes (bittering)

2 HBU (56 MBU) American Willamette hops (pellets)—90 minutes (bittering)

2 HBU (56 MBU) American Cascade hops (pellets)—30 minutes (flavor)

3 HBU (85 MBU) American Willamette hops (pellets)—15 minutes (flavor)

¼ tsp. Irish moss
¾ c. corn sugar for priming in bottles. Use ⅓ cup corn sugar if priming a keg.
Wyeast 1056 American Ale yeast

A single-step infusion mash is employed to mash the grains. Add 13 quarts (12.4 l.) of 169-degree F (76 C) water to the crushed grain, stir, stabilize and hold the temperature at 152 degrees F (66.5 C) for 60 minutes.

After conversion, raise temperature to 167 degrees F (75 C), lauter and sparge with 4.5 gallons (15 l.) of 170-degree F (77 C) water. Collect about 6 gallons (22.8 l.) of runoff, add bittering hops and bring to a full and vigorous boil.

The total boil time will be 90 minutes. When 30 minutes remain, add flavor hops. When 15 minutes remain, add aroma hops and Irish moss. After a total wort boil of 90 minutes (reducing the wort volume to just over 5 gallons), turn off the heat, then separate or strain out and sparge hops. Chill the wort to 65 to 70 degrees F (18–21 C) and direct into a sanitized fermenter. Aerate the cooled wort well. Add an active yeast culture and ferment for 4 to 6 days in the primary. Then transfer into a secondary fermenter, chill to 55 to 60 degrees F (13–15.5 C), and let secondary finish fermentation and cold aging for another two weeks.

When secondary aging is complete, prime with sugar, bottle or keg. Let condition at temperatures above 60 degrees F (15.5 C) until clear and carbonated.

MASH-EXTRACT RECIPE AND PROCEDURE FOR ZOSER OATMEAL STOUT

5	lbs.	(2.3 kg.)	English dark dried malt extract
2	lbs.	(0.9 kg.)	American 2-row pale malt
½	lb.	(0.23 kg.)	American roasted barley
1	lb.	(0.45 kg.)	American caramel malt—130 Lovibond
¾	lb.	(0.34 kg.)	oatmeal
¾	lb.	(0.34 kg.)	American chocolate malt

¼   lb.   (114 g.)   American black malt
5¼   lbs.   (2.4 kg.)   Total grains

3.5   HBU   Cascade hops (pellets)—90 minutes (bittering)
3   HBU   (85 MBU) American Willamette hops (pellets)—90 minutes (bittering)
2   HBU   (56 MBU) American Cascade hops (pellets)—30 minutes (flavor)
3   HBU   (85 MBU) American Willamette hops (pellets)—15 minutes (flavor)

¼   tsp.   Irish moss
¾   c.   corn sugar for priming in bottles. Use ⅓ cup corn sugar if priming a keg.
Wyeast 1056 American Ale yeast

A single-step infusion mash is employed to mash the grains. Add 5 quarts (4.8 l.) of 169-degree F (76 C) water to the crushed grain, stir, stabilize and hold the temperature at 152 degrees F (66.5 C) for 60 minutes.

After conversion, raise temperature to 167 degrees F (75 C), lauter and sparge runoff with 2.5 gallons (9.5 l.) of 170-degree F (77 C) water. Collect about 3 gallons (11.5 l.) of runoff. Add malt extract and bittering hops, and bring to a full and vigorous boil.

The total boil time will be 75 minutes. When 30 minutes remain, add flavor hops. When 15 minutes remain, add aroma hops and Irish moss. After a total wort boil of 75 minutes (reducing the wort volume to just over 5 gallons), turn off the heat. Then separate or strain out and sparge hops, and direct the hot wort into a sanitized fermenter to which 2 gallons (7.6 l.) of cold water have been added. If necessary, add additional cold water to achieve a 5-gallon (19–l.) batch size. Chill the wort to 70 degrees F (21 C). Aerate the cooled wort very well. Add a healthy and very active yeast culture and ferment for 4 to 6 days in the primary. Then transfer into a secondary fermenter, chill to 55 to 60 degrees F (13–15.5 C), and let secondary finish fermentation and cold aging for another two weeks.

When secondary aging is complete, prime with sugar, bottle or keg. Let condition at temperatures above 60 degrees F (15.5 C) until clear and carbonated.

Silver Cup Winner

None

Bronze Cup Winner

Gray's Classic Oatmeal Stout
Gray Brewing Co.
Janesville, Wisconsin, USA

If you take the time to observe, you'll see a hint of evocative ruby red peering through the edges of this dark but not black stout. Malt aroma is a combination of darkly roasted chocolate malts along with aromatic biscuitlike malt and caramel malt. Aromatics of roast barley are strong and expressive, like black coffee with a slightly burnt/phenolic character in a pleasant way. Flavor revisits the strong roasted barley character, while bitterness is not particularly high, though well balanced with specialty-malt sweetness, cocoa and light astringent bitterness of roasted barley. Oatmeal enhances the smooth medium body and overall texture. Medium body is soothed with the texture that oatmeal brings to a well-made stout. No hop flavor or aroma to speak of.

This beer won a Gold Medal in the 1994 GABF in the Specialty Stout category.

Estimated profile based on tasting
Color: 26+ SRM (52+ EBC)
Bittering Units: 28–30

CATEGORY 15: IMPERIAL STOUT

Dark copper to very black, Imperial Stouts typically have
alcohol contents exceeding 8 percent. The extremely rich
malty flavor and aroma are balanced with assertive hopping
and fruity-ester character. Perceived bitterness can be moderate and in balance with malt character to very high in the
darker versions. Roasted-malt astringency and bitterness can
be moderately perceived but should not overwhelm the overall character. Hop aroma can be subtle to overwhelmingly floral. Diacetyl (butterscotch) levels should be very low.

Original Gravity (°Plato):	1.075–1.090 (19–22.5 °Plato)
Apparent Extract– Final Gravity (°Plato):	1.020–1.030 (4–7.5 °Plato)
Alcohol by weight (volume):	5.5–7% (7–9%)
Bitterness (IBU):	50–80
Color SRM (EBC):	20+ (80+ EBC)

Gold Cup Winner

Old Rasputin Russian Imperial Stout
North Coast Brewing Co.
444 North Main St.
Fort Bragg, California, USA 95437
Brewmaster: Mark Ruedrich
Established 1988
Production: 9,000 bbl. (10,500 hl.)

Homebrew. It endears and captures imaginations. Mark
Ruedrich honed his skills as a homebrewer while living in
England in the late 1970s. There, he discovered what he
refers to as "real beer" and soon became involved with the
Campaign for Real Ale (CAMRA). Moving back to the United

Mark E. Ruedrich

Mark Ruedrich, Brewmaster

States and settling in Northern California, he encountered homebrew destiny when he met his future partners, Tom Allen and Joe Rosenthal. Destiny proceeded while he worked construction with Tom for Joe's nineteenth-century inn. The combination of brewing, restaurant management and construction skills evolved along with home fermentations. Through an all-natural progression of events and hard work, the three partners converted a building that at various times had been a mortuary, a church and a college into an award-winning brewery and restaurant recently named Mendocino County Restaurant of the Year.

Situated on the rugged California coastline, the brewery is complemented by towering redwoods and fishermen's wharves, offering comforting ambience, warmth and *mucho gusto*.

North Coast offers several ales and lagers, including their Great American Beer Festival award-winning Red Seal Ale, Scrimshaw Pilsner, Old No. 38 Stout, Alt Nouveau and Oktoberfest Ale. Old Rasputin Russian Imperial Stout now joins their repertoire as a Gold Cup winner in the World Beer Cup. Along with other seasonal specialities such as Blue Star Wheat beer, it is available in their taproom and grill.

North Coast Brewing Company says a lot for the destiny that homebrew can inspire; you know, the stuff between the bubbles.

CHARACTER DESCRIPTION OF GOLD CUP–WINNING OLD RASPUTIN RUSSIAN IMPERIAL STOUT

There's no doubt about the darkness—just black with a rich brown head. A fruit berry-strawberry-plumlike aroma is an overture beginning this dark and joyous symphony. If fruitiness can be described as having a high note, I'll take the liberty to describe it as such. Clean with not a hint of initial banana or apple. Then as cellos, the roast malt emerges after initial confrontation with fruitiness. Ahh, then there emerges a brief allusion to banana in the form of a mild taffylike character with hints of caramel, sweet biscuit, rich toffeelike candy aromas. Like silk, the flavor enraptures with a rich and full-bodied consistency. The aftertaste reminisces as dry-roasted malt and hop bitterness. Astoundingly well balanced. All the complex components of fruitiness, hops, bitterness, flavor, aroma, sweetness, malt, astringency and alcohol emerge, stated simply as black silk. The black-malt bitterness is a serious component of flavor, body and texture. The body is full and hop bitterness low enough so that the overall bitterness is a casual but memorable experience. The high alcohol is warming. Each sip makes you feel rich. (Being rich is not about how much beer you have, but how you enjoy the beer you *do* have.) With eyes closed (and that gets easier with each delectable swallow), dreams of your existence become only pleasant ones. A beer to impart a truly spiritual happiness.

- **Recipe for 5 U.S. gallons (19 liters) Old Rasputin Russian Imperial Stout**

Note: because of the high malt content, process efficiency is 70 percent rather than 75 percent.

Targets:
Original Gravity: 1.090 (21.5)
Final Gravity: 1.018 (4.5)
Alcohol by volume: 9.2%
Color: Black SRM (Black EBC)
Bittering Units: Actual at 55, though calculated to 81 with the consideration that density of full-grain wort reduces utilization.

ALL-GRAIN RECIPE AND PRECEDURE

13½	lbs.	(5.7 kg.)	American 2-row Klages or English 2-row Harrington pale malt
1½	lbs.	(0.69 kg.)	English crystal malt—120 Lovibond
¾	lb.	(0.34 kg.)	American Victory or other aromatic malt
1¼	lbs.	(0.57 kg.)	English chocolate malt
1	lb.	(0.45 kg.)	English roasted barley
½	lb.	(0.23 kg.)	English black malt
2½	lbs.	(1.1 kg.)	English brown malt
21	**lbs.**	**(9 kg.)**	**Total grains**

12	HBU	(340 MBU) Cluster hops (pellets)—105 minutes (bittering)
4	HBU	(113 MBU) American Centennial hops (pellets)—105 minutes (bittering)
6	HBU	(170 MBU) American Liberty hops (pellets)—30 minutes (flavor)
4	HBU	(113 MBU) American Northern Brewer hops (pellets)—30 minutes (flavor)

1½   oz.   (42 g.) American Liberty hops (pellets)—dry-
             hop for 2 weeks (aroma)

¼   tsp.   Irish moss
¾   c.   corn sugar for priming in bottles. Use ⅓ cup
             corn sugar if priming a keg.
Wyeast 1272 American Ale II yeast is suggested.

A single-step infusion mash is employed to mash the grains. Add 5 gallons (19 l.) of 172-degree F (78 C) water to the crushed grain, stir, stabilize and hold the temperature at 155 degrees F (68.5 C) for 60 minutes.

After conversion, raise temperature to 167 degrees F (75 C), lauter and sparge with 5 gallons (19 l.) of 170-degree F (77 C) water. Collect about 8 gallons (30 l.) of runoff, add bittering hops and bring to a full and vigorous boil.

The total boil time is estimated at 105 minutes in order to evaporate the wort down to the final batch size of 5 gallons (19 l.). When 30 minutes remain, add flavor hops. When 10 minutes remain, add Irish moss. After a total wort boil of about 105 minutes (reducing the wort volume to just over 5 gallons), turn off the heat, then separate or strain out and sparge hops. Chill the wort to 65 to 70 degrees F (18–21 C) and direct into a sanitized fermenter. Aerate the cooled wort extremely well. Add a very good dose of an active yeast culture and ferment in the primary until gravity drops to 1.032–1.036 (8–9) (an estimated 7 to 8 days in the primary). Then transfer into a secondary fermenter, chill to 60 degrees F (13–15.5 C) if possible. Let age for two weeks, then add dry hops. Then age two additional weeks. Remove dry hops or rack off into another vessel and age for one month.

When aging is complete, prime with sugar, bottle or keg. Let condition at temperatures above 60 degrees F (15.5 C) until clear and carbonated.

MASH-EXTRACT RECIPE AND PROCEDURE FOR OLD RASPUTIN RUSSIAN IMPERIAL STOUT

5¾	lbs.	(2.6 kg.)	English amber dried malt extract
2	lbs.	(0.9 kg.)	American 2-row Klages or English 2-row Harrington pale malt
1½	lbs.	(0.69 kg.)	English crystal malt—120 Lovibond
¾	lb.	(0.34 kg.)	American Victory or other aromatic malt
1¼	lbs.	(0.57 kg.)	English chocolate malt
1	lb.	(0.45 kg.)	English roasted barley
½	lb.	(0.23 kg.)	English black malt
2½	lbs.	(1.1 kg.)	English brown malt
9½	**lbs.**	**(4.3 kg.)**	**Total grains**

14	HBU	(397 MBU) Cluster hops (pellets)—105 minutes (bittering)
5	HBU	(142 MBU) American Centennial hops (pellets)—105 minutes (bittering)
6	HBU	(170 MBU) American Liberty hops (pellets)—30 minutes (flavor)
4	HBU	(113 MBU) American Northern Brewer hops (pellets)—30 minutes (flavor)
1½	oz.	(42 g.) American Liberty hops (pellets)—dry-hop for two weeks (aroma)

¼	tsp.	Irish moss
¾	c.	corn sugar for priming in bottles. Use ⅓ cup corn sugar if priming a keg.

Wyeast 1272 American Ale II yeast is suggested.

A single-step infusion mash is employed to mash the grains. Add 9.5 quarts (9 l.) of 172-degree F (78 C) water to the crushed grain, stir, stabilize and hold the temperature at 155 degrees F (68.5 C) for 60 minutes.

After conversion, raise temperature to 167 degrees F (75 C), lauter and sparge with 4 gallons (15 l.) of 170-degree F (77 C) water. Collect about 5.5 gallons (11.5 l.) of runoff.

Add malt extract and bittering hops and bring to a full and vigorous boil.

The total boil time is estimated at 105 minutes in order to evaporate the wort down to the final concentrated wort of about 3.5 gallons (13 l.) yield. When 30 minutes remain, add flavor hops. When 10 minutes remain, add Irish moss. After a total wort boil of about 105 minutes (reducing the wort volume to about 3.5 gallons [13 l.]), turn off the heat, then separate or strain out and sparge hops, and direct the hot wort into a sanitized fermenter to which 2 gallons (7.6 l.) of cold water have been added. If necessary, add additional cold water to achieve a 5-gallon (19–l.) batch size. Chill the wort to 65–70 degrees F (18.5–21 C). Aerate the cooled wort extremely well. Add a very good dose of an active yeast culture and ferment in the primary until gravity drops to 1.032–1.036 (8–9) (an estimated 7 to 8 days in the primary). Then transfer into a secondary fermenter, chill to 60 degrees F (13–15.5 C) if possible. Let age for two weeks, then add dry hops. Then age two additional weeks. Remove dry hops or rack off into another vessel and age for one month.

When aging is complete, prime with sugar, bottle or keg. Let condition at temperatures above 60 degrees F (15.5 C) until clear and carbonated.

Silver Cup Winner

Imperial Stout
Wiibroes Brewery Co.
Helsingor, Denmark

Rich, light brown head secures a deep, dark black imperial stout. Bubbles languidly rise to the surface, suggesting a full body. Beautifully aromatic mild roast coffee/chocolate malt intensifies as smoothness. Also woven into this rich and malty stout is a fleeting suggestion of floral hops, but one can't be quite sure with such a rich roast-malt character. Flavor impression is immediately alcoholic accompanied by a remarkable clean dryness followed by an amusingly astringent roasted malt and simple balance of hop bitterness. Bitterness is straightforward with no flavor overtones. Full-bodied, well-attenuated and cleanly brewed high-gravity stout (also called Wiibroe Porter). An imperial stout in a classic European sense, it is not complicated with floral and flavor hops, but rather relies on clean bitterness, controlled fermentation and a complexity of roasted malts to create a complementing dryness. The initial impression of the full body continues in the aftertaste in a devious way as one settles down with this refreshingly clean, strong and wonderfully deceiving stout. A very reflective and friendly beer. My thoughts become one with Wiibroe. (Omigosh, I drank the whole bottle. No more tasting today.)

Brewery formulation uses Danish Pilsner, German Colour, Munchener and Karamel malts used with German Hallertauer hops for bitterness. Bottom-fermented with lager yeast.

Estimated profile based on tasting
Original Gravity: 1.081 (19.5) indicated by the brewery
Final Gravity: 1.019 (4.8) indicated by the brewery
Alcohol by volume: 8.2% indicated by the brewery
Color: 30+ SRM (60+ EBC)
Bittering Units: 40–45 (43 indicated by the brewery)

Bronze Cup Winner

Centennial Russian Imperial Stout
Trinity Brewhouse
Providence, Rhode Island, USA

Close your eyes on a moonless night. There are no ales darker. Surprisingly, there seems to be no suggestion of roasted malt, roasted barley or alcohol aroma. But that is my fault for having chilled this stout excessively. As it warms and as any good ale will do, these characters emerge as a lost friend found. Alelike fruitiness blends nicely. Flavor is medium- to full-bodied with a quick, brief caramel sweetness swiftly evolving to a roasted-barley, malt and hop bitterness, sending the senses over the edge. You've got to love it if stout is your style. Bitterness is a real asset, followed by a dry finish, perfect for those who would rather not dwell with sweetness. Aftertaste is a lot of big roast-barley and malt dryness with brief hop bitterness. Clean but boisterous. One would think that alcohol should be making an impression, but like a stage manager, it assures perfect choreography and rules behind the scenes, until later. Good night.

Estimated profile based on tasting
Color: 1000+ SRM (2000+ EBC)
Bittering Units: 50–60

Irish Origin

CATEGORY 16:
CLASSIC IRISH-STYLE DRY STOUT

Dry Stouts have an initial malt and caramel flavor profile with a distinctive dry-roasted bitterness in the finish. Dry Stouts achieve a dry-roasted character from the use of roasted barley. Some slight acidity may be perceived but is not necessary. Hop aroma and flavor should not be perceived. It has a medium body. Fruity esters are minimal and are overshadowed by malt, hop bitterness and roasted-barley character. Diacetyl (butterscotch) should be very low or not perceived. Head retention and rich character should be part of its visual character.

Original Gravity (°Plato):   1.038–1.048 (9.5–12 °Plato)
Apparent Extract–
Final Gravity (°Plato):   1.008–1.014 (2–3.5 °Plato)

Alcohol by weight (volume):   3.2–4.2% (3.8–5%)
Bitterness (IBU):   30–40
Color (SRM):   40+ (150+ EBC)

Gold Cup Winner

Founders Stout
Mishawaka Brewing Company
3703 North Main Street
Mishawaka, Indiana, USA 46545
Brewmasters: Tom and Rick Schmidt
Established 1992
Production: 1,300 bbl. (1,500 hl.)

Asked about the brewery's great moment, Tom Schmidt replies, "Any day has its 'great moments' if one is to sit down to a slab of our 'Oh Baby! Oh Baby Back Ribs' and a pint of any of our beers." Great moments seem to be what it's

Tom Schmidt and Rick Schmidt, Brewmasters

Thomas R. Schmidt

all about at the Mishawaka Brewing Company in the fabled football area of South Bend, Indiana. On June 14, 1996, the brewpub received a postcard from regular patrons Jim and Sharon Turley. The card, written as they drank a luscious glass of Guinness in a small pub in Ireland, noted, "We both agree that as good as the Guinness is here on the 'old sod,' we still like your Founders Stout better!" The next day the brewery learned their Founders Stout had won the Gold Cup at the World Beer Cup.

Fascinated by the enthusiasm and stories of success they heard from attendees at the first National Microbrewers Conference in 1986 (now called the National Craft Brewers Conference and Trade Show), Tom Schmidt and Jim Foster began their journey to brewpub reality. Years of research, brewery visits, legal maneuvering and careful planning culminated in the 1992 brewery and brewpub opening. The city of Mishawaka had its first brewery. The 9,000-square-foot building, formerly housing a fitness center, became a "fitness" center of another sort with a 330-seat restaurant and outdoor beer garden.

The brewery routinely uses three strains of yeast for its ales, lagers and wheat beers. Interestingly, the brewery has discovered that its patrons prefer a degree of diacetyl (a butterscotch character); consequently, it pays careful attention to the temperature during fermentation to achieve the desired character in many of the brews.

One of the problems you'll find upon a visit to the brewpub is having to choose from the tremendous variety of beers on tap. Four Horseman Ale, Lake Effect Pale Ale, INDIAna Pale Ale, Silver Hawks Pilsner, Ankenbrock Weizen, Raspberry Wheat and Gold Cup winner Founders Stout are almost always available year-round. Specialty beers are brewed, such as Grumpy's Oatmeal Stout, Hop Head Ale, kNUTe Brown Ale, Oktoberfest Beer, Resolution (barley wine) Ale, Cherry Wheat, Dominator Doppelbock and Jac 'O' Lager Pumpkin Beer.

This introduction wouldn't be complete without mention

of their homebrew connection. With extensive backgrounds in manufacturing, science, sales and marketing, Foster and Schmidt already had some beery ideas percolating by early 1987. Tom Schmidt constructed a pilot-scale brewery for his own self-taught home course in the art and science of brewing. It's paid off with great beer and international recognition, both at the World Beer Cup and the Great American Beer Festival.

CHARACTER DESCRIPTION OF GOLD CUP–WINNING FOUNDERS STOUT

Simply rich, seductive and exotically black. There is nothing transparent about this brew. Though the label claims this beer is made according to the Reinheitsgebot, fortunately it is not. The rich nonmalted roasted barley is a preciously important ingredient in this brew and is actually not allowed according to the German purity law, which states that beer can only be made with malt, water, hops and modern-day yeast. The aroma expresses a wonderful underlying complex alelike fruitiness, a caramel-toffeelike richness and a chocolate-cocoa character that is evident but subtle. And ah, yes, the wonderful expression of roasted barley proliferates. Overall a clean, sweet aroma with suggestions of dark caramel malt (expressed as a light sweetness). Smooth and velvetlike flavor indicating a cocoa character, with the first impression a brief burst of roasted malt and sweetness followed by a smooth transition to hop bitterness. A relatively light body creates the impression of a wonderfully refreshing beer, contrary to its rich and seductive appearance. While not Guinness (which wasn't entered into this competition), this beer has more character than Guinness in its flavor profile. Surprisingly, the aftertaste is not bitter but relatively clean and neutral, with just enough roast-malt, roast-barley and hop bitterness left to complement the overall impression. There is a notable alcoholic tingle in the aftertaste. It's a stout that is so complex it leaves you wondering about your overall impression; a seductive coffeelike bitterness lingers. It is extremely well balanced. No hop aroma or flavor.

Goldings would be a good choice of hops. Even American Goldings would work as the character of English Goldings would be lost. The brewery's recipe dry-hops this stout, yet it is so well integrated it is lost in the overall complexity.

- ## Recipe for 5 U.S. gallons (19 liters) Founders Stout
  Targets:
  Original Gravity: 1.052 (13)
  Final Gravity: 1.016 (4)
  Alcohol by volume: 4.6%
  Color: 56 Black SRM (122 EBC)
  Bittering Units: 37

ALL-GRAIN RECIPE AND PROCEDURE

7	lbs.	(3.2 kg.)	American 2-row pale malt
1½	lbs.	(0.68 kg.)	American caramel malt—20 Lovibond
½	lb.	(0.23 kg.)	American chocolate malt
½	lb.	(0.23 kg.)	American roasted barley
½	lb.	(0.23 kg.)	American stout roast
¼	lb.	(114 g.)	American black malt
10¼	**lbs.**	**(4.7 kg.)**	**Total grains**

3.5 HBU (99 MBU) American Perle hops (pellets)—90 minutes (bittering)

3.5 HBU (99 MBU) American Willamette hops (pellets)—90 minutes (bittering)

2.5 HBU (71 MBU) English Kent Goldings hops (pellets)—15 minutes (flavor)

¼ oz. (7 g.) American Mt. Hood hops (pellets)—dry-hop for 2 weeks (aroma)

¼   tsp.   Irish moss
¾   c.   corn sugar for priming in bottles. Use ⅓ cup
                corn sugar if priming a keg.
Wyeast 1335 British Ale yeast or Wyeast 1098 British Ale
yeast suggested.

A single-step infusion mash is employed to mash the
grains. Add 10 quarts (9.5 l.) of 170-degree F (77 C) water
to the crushed grain, stir, stabilize and hold the temperature
at 153 degrees F (67.5 C) for 60 minutes.

After conversion, raise temperature to 167 degrees F (75
C), lauter and sparge with 5 gallons (15 l.) of 170-degree F
(77 C) water. Collect about 6 gallons (23 l.) of runoff, add
bittering hops and bring to a full and vigorous boil.

The total boil time will be 90 minutes. When 15 minutes
remain, add flavor hops and Irish moss. After a total wort
boil of 90 minutes (reducing the wort volume to just over
5 gallons), turn off the heat, then separate or strain out and
sparge hops. Chill the wort to 65 to 70 degrees F (18–21 C)
and direct into a sanitized fermenter. Aerate the cooled wort
well. Add an active yeast culture and ferment for 4 to 6 days
in the primary. Then transfer into a secondary fermenter and
add aroma-dry hops. Allow to age for two weeks or more.

When secondary aging is complete, prime with sugar, bottle or keg. Let condition at temperatures above 60 degrees
F (15.5 C) until clear and carbonated.

MALT-EXTRACT RECIPE AND PROCEDURE FOR FOUNDERS STOUT

4¾   lbs.   (2.2 kg.)   English dark dried malt extract
1½   lbs.   (0.68 kg.)   American caramel malt—20 Lovi-
                          bond
½   lb.   (0.23 kg.)   American chocolate malt
¼   lb.   (114 g.)   American roasted barley
½   lb.   (0.23 kg.)   American stout roast
2¾   lbs.   (1.2 kg.)   Total grains

5.5	HBU	(156 MBU) American Perle hops (pellets)—60 minutes (bittering)
4	HBU	(113 MBU) American Willamette hops (pellets)—60 minutes (bittering)
2.5	HBU	(71 MBU) English Kent Goldings hops (pellets)—15 minutes (flavor)
¼	oz.	(7 g.) American Mt. Hood hops (pellets)—dry-hop for 2 weeks (aroma)
¼	tsp.	Irish moss
¾	c.	corn sugar for priming in bottles. Use ⅓ cup corn sugar if priming a keg.

Wyeast 1335 British Ale yeast or Wyeast 1098 British Ale yeast suggested.

Steep crushed specialty grains in 1½ gallons (5.7 l.) water at 150 degrees F (65.5 C) for 30 minutes. Strain and sparge with enough 170-degree F (76.5 C) water to finish with 2½ gallons (9.5 l.) specialty grain liquor. Add the dried malt extract and bittering hops and bring to a full and vigorous boil.

The total boil time will be 60 minutes. When 15 minutes remain, add flavor hops and Irish moss. After a total wort boil of 60 minutes, turn off the heat, separate or strain out and sparge hops, and direct the hot wort into a sanitized fermenter to which 2 gallons (7.6 l.) of cold water have been added. If necessary, add additional cold water to achieve a 5-gallon (19–l.) batch size. Chill the wort to 70 degrees F (21 C). Aerate the cooled wort well. Add an active yeast culture and ferment for 4 to 6 days in the primary. Then transfer into a secondary fermenter and add aroma-dry hops. Allow to age for two weeks or more.

When secondary aging is complete, prime with sugar, bottle or keg. Let condition at temperatures above 60 degrees F (15.5 C) until clear and carbonated.

Silver Cup Winner

Neptune Black Sea Stout
Neptune Brewery Co.
New York, New York, USA

Dark as a stout should be. Roast-coffee character evolves in the aroma without any hop character, as it should for style. Sweet caramel is evident but not overdone; in fact, its suggestive presence serves only to balance the robust roast-barley and roast-malt character. Impressively light- to medium-bodied stout with good clean malt notes balanced by the sharp bite of hop bitterness and the astringent character of roast barley. A slight cocoa flavor dwells playfully in the background, but it is the roast barley–coffee character that impresses most. Floral hop flavor comes through after the initial pleasantries of malt assault. Dry, bitter aftertaste is like a good clean cup of black coffee, but with the additional pleasant complexity of floral hop flavor, it's more than any cup of java will ever be.

Estimated profile based on tasting
Color: 30+ SRM (60+ EBC)
Bittering Units: 37–41

Bronze Cup Winner

Seminole Stout
Buckhead Brewery and Grill
Tallahassee, Florida, USA

Deeply dark and opaque stout. Very clean aroma with appropriate absence of hop character. Foundation for aroma is an earthy, subtle roasted-barley character. Slightly smoky and coffeelike, but not overstated. Intriguing foundation. Mouth feel is medium-bodied (almost full-bodied) with a flavor based on roasted barley and caramel malt. Some bit-terness, certainly attributable to the roasted malts and bar-ley. Hop bitterness is clean and definitive, being neither gentle nor soft, but rather punctual. Aftertaste is slightly reminiscent of caramel malt, but roasted and hop bitterness carry the day. Quite a refreshing stout with clean attributes that serve to help define and represent this style well. Malt, roasted barley and hops say it all.

Brewery formulation uses American dark crystal, choco-late, black, Munich and wheat malts, roasted barley and oats. Northern Brewer and Willamette hops are used for bitterness and flavor.

Estimated profile based on tasting
Color: 30+ SRM (60+ EBC)
Bittering Units: 36–40

CATEGORY 17: FOREIGN-STYLE STOUT

As with Classic Dry Stouts, Foreign-Style Stouts have an initial malt sweetness and caramel flavor with a distinctive dry-roasted bitterness in the finish. Some slight acidity is permissible and a medium- to full-bodied mouth feel is ap-propriate. Hop aroma and flavor should not be perceived. The perception of fruity esters is low. Diacetyl (butterscotch) should be negligible or not perceived. Head retention is excellent.

Original Gravity (°Plato):	1.052–1.072 (13–18 °Plato)
Apparent Extract—	
Final Gravity (°Plato):	1.008–1.020 (2–5 °Plato)
Alcohol by weight (volume):	4.8–6% (6–7.5%)
Bitterness (IBU):	30–60
Color (SRM):	40+ (150+ EBC)

Gold Cup Winner

San Quentin's Breakout Stout

Marin Brewing Company
1809 Larkspur Landing Circle
Larkspur, California, USA 94939
Brewmasters: Brendan J. Moylan and Arne Johnson
Established 1989
Production: 2,700 bbl. (3,200 hl.)

San Quentin also won the Strong Ale Category. Refer to that category for a description of the brewery.

CHARACTER DESCRIPTION OF GOLD CUP–WINNING SAN QUENTIN'S BREAKOUT STOUT

With an erotic glimmer, a faint red hue barely pierces through this black pearl, covered with a clean white head of foam. Chocolate and roast malt blast out in the aroma. Fruity esters are subliminally low with a suggestion of caramel. Hop aroma is not evident. With a first taste one is met

Brendan J. Moylan

with the impression of a medium sweet and medium-bodied stout extraordinarily balanced with hop bitterness. Finely tuned and artfully brewed. The roast-malt character does not contribute to any astringent bitterness or mouth feel. A second and third mouthful still portends a medium-bodied beer, but with great attenuation. Malt sweetness feels like it is a result of high mashing temperatures rather than a lot of caramel. The sweet and bitter balance so important in a stout is dead center with a slight bitter hop aftertaste. Hops do not portray flavor, only a soft, clean bitterness. Generally aftertaste is mildly bitter with a roast-malt character after an initial hit of sweet and full malt sweetness. An ample brew for ample times. Quenching balance promotes drinkability.

• *Recipe for 5 U.S. gallons (19 liters) San Quentin's Breakout Stout*

Targets:
Original Gravity: 1.065 (16)
Final Gravity: 1.020 (5)
Alcohol by volume: 5.8%
Color: 84 SRM (168 EBC)
Bittering Units: 28

ALL-GRAIN RECIPE AND PROCEDURE

10¼ lbs.	(4.7 kg.)	American 2-row Klages/Harrington pale malt
½ lb.	(0.23 kg.)	English crystal malt—80 Lovibond
½ lb.	(0.23 kg.)	flaked barley
1 lb.	(0.45 kg.)	American roasted barley
½ lb.	(0.23 kg.)	American black malt
12¾ lbs.	**(5.8 kg.)**	**Total grains**

4 HBU (113 MBU) American Chinook hops (whole)—105 minutes (bittering)

3.2 HBU (91 MBU) English Northdown hops (whole)—105 minutes (bittering)

¼ tsp. Irish moss
¾ c. corn sugar for priming in bottles. Use ⅓ cup corn sugar if priming a keg.

Wyeast 1968 London ESB Ale yeast

A single-step infusion mash is employed to mash the grains. Add 13 quarts (12 l.) of 175-degree F (80 C) water to the crushed grain, stir, stabilize and hold the temperature at 158 degrees F (70 C) for 60 minutes.

After conversion, raise temperature to 167 degrees F (75 C), lauter and sparge with 4.5 gallons (17 l.) of 170-degree F (77 C) water. Collect about 6 gallons (23 l.) of runoff, add bittering hops and bring to a full and vigorous boil.

The total boil time will be 105 minutes. When 10 minutes remain, add Irish moss. After a total wort boil of 105 minutes (reducing the wort volume to just over 5 gallons), turn off the heat. Then separate or strain out and sparge hops. Chill the wort to 65 to 70 degrees F (18–21 C) and direct into a sanitized fermenter. Aerate the cooled wort well. Add an active yeast culture and ferment for 4 to 6 days in the primary. Then transfer into a secondary fermenter and age for two to three weeks.

When secondary aging is complete, prime with sugar, bottle or keg. Let condition at temperatures above 60 degrees F (15.5 C) until clear and carbonated.

MASH-EXTRACT RECIPE AND PROCEDURE FOR SAN QUENTIN'S BREAKOUT STOUT

4¾	lbs. (2.2 kg.)	English dark dried malt extract
2½	lbs. (1.1 kg.)	American 2-row Klages/Harrington pale malt
½	lb. (0.23 kg.)	English crystal malt—80 Lovibond
½	lb. (0.23 kg.)	flaked barley
1	lb. (0.45 kg.)	American roasted barley
½	lb. (0.23. kg.)	American black malt
5	**lbs. (2.3 kg.)**	**Total grains**
5	HBU	(142 MBU) American Chinook hops (whole)—60 minutes (bittering)
5	HBU	(142 MBU) English Northdown hops (whole)—60 minutes (bittering)

¼ tsp. Irish moss
¾ c. corn sugar for priming in bottles. Use ⅓ cup corn sugar if priming a keg.
Wyeast 1968 London ESB Ale yeast

A single-step infusion mash is employed to mash the grains. Add 5 quarts (4.8 l.) of 175-degree F (80 C) water to the crushed grain, stir, stabilize and hold the temperature at 158 degrees F (70 C) for 60 minutes.

After conversion, raise temperature to 167 degrees F (75 C), lauter and sparge with 2 gallons (7.6 l.) of 170-degree F (77 C) water. Add water if necessary to collect about 2.5 gallons (9.5 l.) of runoff. Add malt extract and bittering hops and bring to a full and vigorous boil.

The total boil time will be 60 minutes. When 10 minutes remain, add Irish moss. After a total wort boil of 60 minutes (reducing the wort volume to just over 5 gallons), turn off the heat. Then separate or strain out and sparge hops, and direct the hot wort into a sanitized fermenter to which 2 gallons (7.6 l.) of cold water have been added. If necessary, add additional cold water to achieve a 5-gallon (19-l.) batch size. Chill the wort to 65 to 70 degrees F (18–21 C). Aerate the cooled wort well. Add an active yeast culture and ferment for 4 to 6 days in the primary. Then transfer into a secondary fermenter and age for two to three weeks.

When secondary aging is complete, prime with sugar, bottle or keg. Let condition at temperatures above 60 degrees F (15.5 C) until clear and carbonated.

Silver Cup Winner

Cascade Special Stout
Cascade Brewery Co.
Hobart, Tasmania, Australia

Rich, dark and alluring color with a suggestion of transparency. Rich roasted malt and unique fruity aroma suggest a dry character. Roast character is vaguely coffeelike with a light aromatic sweetness; not at all chocolatelike. First flavor impression is simply delicious, medium-bodied with an extraordinary clean finish. It has the complex nature of alelike fruitiness, but what really makes this stout distinctive is the toffeelike character of the roasted malt or barley. It is worthy for any brewer to experience and quenching to any stout enthusiast. Hop bitterness is evident, but roast barley/malt obviously contributes to the sensation of bitterness without astringency.

Estimated profile based on tasting
Color: 28+ SRM (56+ EBC)
Bittering Units: 30–36

Bronze Cup Winner

Echigo Stout
Uehara Shuzou, Echigo Beer
Nishikanbara-Gun, Japan

It's a glass of nighttime, extraordinarily black. Wonderfully pleasant, complex fruitiness subtly reminiscent of wild cherries or plums. As the beer warms, the roast malts emerge through the fruitiness in a delicate way. Medium- to full-bodied beer. The intriguing foreplay of a woody, cedarlike smokiness accents the emerging roast-barley flavor. It intoxicates with a sense of anticipation. Finishes extraordinarily dry and clean. A subtle roast-burnt-malt bitterness is sensed in the aftertaste, but the overall sensation is of a slight hop bitterness without complicated interference from hop flavors and malts. This is a well-stated, simple stout with exotic overtures. The brewer has managed to combine simplicity with the poetry of malt and hops. An exceptional example of this style.

As the beer warms, malt and hops begin to emerge with

greater impact, but remain well balanced, assuring one's continued well-being.

Estimated profile based on tasting
Color: 50+ SRM (100+ EBC)
Bittering Units: 36–40

North American Origin

CATEGORY 18:
AMERICAN-STYLE PALE ALE

American Pale Ales range from golden to light copper-colored. The style is characterized by American-variety hops used to produce high hop bitterness, flavor and aroma. American Pale Ales have medium body and low to medium maltiness. Low caramel character is allowable. Fruity-ester flavor and aroma moderate to strong. Diacetyl should be absent or at very low levels. Chill haze is allowable at cold temperatures.

Original Gravity (°Plato):   1.044–1.056 (11–14 °Plato)
Apparent Extract–
Final Gravity (°Plato):   1.008–1.016 (2–4 °Plato)
Alcohol by weight (volume): 3.5–4.3% (4.5–5.5%)

Bitterness (IBU):   20–40
Color SRM (EBC):   4–11 (10–25 EBC)

Gold Cup Winner

Snake River Pale Ale
Snake River Brewing Company
265 S. Millward
Jackson Hole, Wyoming, USA 83001
Brewmaster: Curtis "Chip" Holland
Established 1993
Production: 3,500 bbl. (4,100 hl.)

"What the world needs now is not another beer, but a better beer," reads the mission statement of Wyoming's first brewpub and the first brewery to open in the state since 1954. First established by co-owners Albert and Joni Upsher as the Jackson Hole Pub & Brewery, the brewery was renamed Snake River Brewing Co. Formerly an Anheuser-Busch dis-

Michael Jackson, beer expert and author, and Curtis "Chip" Holland, Brewmaster

Paula Currie

tributor in Oregon for eleven years until 1990, Albert Upsher has accomplished his dream of producing high-quality beers in accordance with the German Purity Laws (malt, hops, water and yeast). Now you'll find six beers on tap at the brewpub in addition to rotating seasonal beers. Beers are brewed 15 barrels at a time, then served from tanks at the brewpub or kegged and bottled for limited distribution in about twelve states.

Besides the Gold Cup–winning Snake River Pale Ale (also a Silver Medal winner in the American Pale Ale category at the 1995 Great American Beer Festival), you'll find Snake River Zonker Stout (Silver Medal winner in the Dry Stout category at the 1994 Great American Beer Festival and Gold Medal at the 1995 Great American Beer Festival), Snake River Alpine Fest (Oktoberfest-style lager), Snake River Celebration (Scottish-style) Ale, Snake River Bald Eagle Bock, Snake River Lager (amber lager), Snake River Brown Ale (a.k.a. Buffalo Brown), Smoked Lager (German Rauchbier style) and Pale Morning Lager (European-style Pils).

CHARACTER DESCRIPTION OF GOLD CUP–WINNING SNAKE RIVER PALE ALE

A standout with crystal clarity, a light golden color and an extremely appealing rich, creamy head clinging as lace on the sides of my glass. Malt sneaks through at secondary levels in the aroma. It is difficult to perceive if the beer is served well chilled. What impresses in the aroma is the soft blend of citruslike American Chinook-Centennial-Cascade hops with an intimation of English Kent Goldings due to the low-level presence of diacetyl (butterscotch character).

Clean, dry, low- to medium-bodied mouth feel is complemented by a well-balanced medium level of soft, citrusy hop flavor and bitterness; assertive but not assaulting. The suggestion of Kent Goldings is evident in the flavor, but is unlikely in the original brewery formulation. Fruitiness of ale is present, contributing to the overall balance of this American-style pale ale.

Overall distinctiveness is attributable to an assertive hop character that does not blast the palate. It seems this glass of beer I'm enjoying certainly deserves the label of a quintessential American pale ale. Aftertaste is relatively clean with a slight hop tang lingering briefly. Malt character lacks any hint of caramel. Its fullness takes on a secondary role in overall impression of character. It's difficult to find any flaws in this beer—and I don't want to.

* *Recipe for 5 U.S. gallons (19 liters) Snake River Pale Ale*

Targets:
Original Gravity: 1.050 (12.5)
Final Gravity: 1.006 (1.5)
Alcohol by volume: 5.8%
Color: 5 SRM (10 EBC)
Bittering Units: 33

ALL-GRAIN RECIPE AND PROCEDURE

8½ lbs.	(3.9 kg.)	American 2-row Klages/Harrington pale malt
¼ lb.	(114 g.)	American caramel malt—20 Lovibond
½ lb.	(0.23 kg.)	American Munich malt—7 Lovibond
9¼ lbs.	**(4.2 kg.)**	**Total grains**

7	HBU	(198 MBU) American Chinook hops (whole)—90 minutes (bittering)
1	HBU	(28 MBU) English Kent Goldings hops (whole)—30 minutes (flavor)
½	oz.	(14 g.) American Cascade hops (whole)—steep in finished boiled wort for 2 to 3 minutes (aroma)
½	oz.	(14 g.) American Cascade hops (whole)—dry-hop for 2 weeks (aroma)

¼   tsp.   Irish moss
¾   c.   corn sugar for priming in bottles. Use ⅓ cup
              corn sugar if priming a keg.
Wyeast 1056 American Ale yeast recommended

A step infusion mash is employed to mash the grains. Add 9 quarts (8.6 l.) of 136-degree F (58 C) water to the crushed grain, stir, stabilize and hold the temperature at 128 degrees F (53 C) for 30 minutes. Add 5 quarts (4.8 l.) of boiling water. Add heat to bring temperature up to 150 degrees F (65.5 C). Hold for 60 minutes.

After conversion, raise temperature to 167 degrees F (75 C), lauter and sparge with 4 gallons (19 l.) of 170-degree F (77 C) water. Collect about 6 gallons (23 l.) of runoff, add bittering hops and bring to a full and vigorous boil.

The total boil time will be 90 minutes. When 30 minutes remain, add flavor hops. When 10 minutes remain, add Irish moss. After a total wort boil of 90 minutes (reducing the wort volume to just over 5 gallons), turn off the heat, add aroma hops and let steep 2 to 5 minutes. Then separate or strain out and sparge hops. Chill the wort to 65 to 70 degrees F (18–21 C) and direct into a sanitized fermenter. Aerate the cooled wort well. Add an active yeast culture and ferment for 4 to 6 days in the primary. Then transfer into a secondary fermenter, chill to 60 degrees F (13–15.5 C) if possible and age for two weeks. Then add aroma-dry hops. Allow to age for two weeks or more.

When aging is complete, prime with sugar, bottle or keg. Let condition at temperatures above 60 degrees F (15.5 C) until clear and carbonated.

MALT-EXTRACT RECIPE AND PROCEDURE FOR SNAKE RIVER PALE ALE

5   lbs.   (2.3 kg.)   English light dried malt extract
½   lb.   (0.23 kg.)   English amber dried malt extract
¼   lb.   (114 g.)   American caramel malt—20 Lovibond
¾   lb.   (0.34 kg.)   Total grains

9   HBU   American Chinook hops (whole)—60 minutes (bittering)

1   HBU   (28 MBU) English Kent Goldings hops (whole)—30 minutes (flavor)

½   oz.   (14 g.) American Cascade hops (whole)—steep in finished boiled wort for 2 to 3 minutes (aroma)

½   oz.   (14 g.) American Cascade hops (whole)—dry-hop for 2 weeks (aroma)

¼   tsp.   Irish moss

¾   c.   corn sugar for priming in bottles. Use ⅓ cup corn sugar if priming a keg.

Wyeast 1056 American Ale yeast recommended

Steep crushed specialty grains in 1½ gallons (5.7 l.) water at 150 degrees F (65.5 C) for 30 minutes. Strain and sparge with enough 170-degree F (76.5 C) water to finish with a little over 2.5 gallons (9.5 l.) specialty grain liquor. Add the dried malt extract and bittering hops and bring to a full and vigorous boil.

The total boil time will be 60 minutes. When 30 minutes remain, add flavor hops. When 10 minutes remain, add Irish moss. After a total wort boil of 60 minutes (reducing the wort volume to just over 5 gallons), turn off the heat, add aroma hops and let steep 2 to 5 minutes. Then separate or strain out and sparge hops, and direct the hot wort into a sanitized fermenter to which 2 gallons (7.6 l.) of cold water have been added. If necessary, add additional cold water to achieve a 5-gallon (19–l.) batch size. Chill the wort to 65–70 degrees F (18.5–21 C). Aerate the cooled wort well. Add an active yeast culture and ferment for 4 to 6 days in the primary. Then transfer into a secondary fermenter, chill to 60 degrees F (13–15.5 C) if possible and age for two weeks. Then add aroma-dry hops. Allow to age for two weeks or more.

When aging is complete, prime with sugar, bottle or keg.

Let condition at temperatures above 60 degrees F (15.5 C) until clear and carbonated.

Silver Cup Winner

South Platte Pale Ale
Columbine Mill Brewery Co.
Littleton, Colorado, USA

Light golden amber with an orange hue. American hop along with a malty caramel-like sweetness is evident in aroma, but the symphony is primarily hops. American Cascade-like, citrusy, fruity aroma with a hint of grassiness. Flavor is exceptionally clean and fermentation character neutral. An imposing bitter hop bite is joined by the fruity, citrus hop flavor of Cascade/Centennial type. Medium body fades to a very dry finish, due to the clean bitterness.

Brewery formulation uses English Munich and Belgian crystal malts with Nugget and Cascade hops for bitterness, flavor and aroma.

Estimated profile based on tasting
Original Gravity: 1.050 (12.5) indicated by brewery
Final Gravity: 1.010 (2.5) indicated by brewery
Alcohol by volume: 5% indicated by brewery
Color: 6–7 SRM (12–14 EBC) (15 EBC indicated by the brewery)
Bittering Units: 40–43 (35 indicated by the brewery)

Bronze Cup Winner

Red Rocket Pale Ale
Bristol Brewing Co.
Colorado Springs, Colorado, USA

Amber ale with orange hue. A blend of caramel and toasted malt with citrusy American-style hopping represent the primary aroma. Malt lends support to a medium- to light-bodied mouth feel followed by a clean, dry finish. Hop flavor and soft bitterness in the classic American-grown Cascade/Centennial tradition permeate but do not overrun the senses. Nicely accomplished balance entices rather than assaults. Pleasant alelike fruitiness verifies ale fermentation. Extraordinarily clean finish, with a small lingering bitterness beckoning a return for another. And another and another.

Brewery formulation uses dark crystals, Cara-Pils, Vienna and Victory malts with American Perle, Willamette and Cascade hops for bitterness, flavor and dry-hopping.

Estimated profile based on tasting
Original Gravity: 1.050 (12.5) indicated by the brewery
Final Gravity: 1.014 (3.5) indicated by the brewery
Color: 12–14 SRM (24–28 EBC)
Bittering Units: 35–37 (30–32 indicated by the brewery)

CATEGORY 19:
AMERICAN-STYLE AMBER ALE

American Amber Ales range from light copper to light brown in color. Amber Ales are characterized by American varietal hops used to produce high hop bitterness, flavor and aroma. Amber Ales have medium to high maltiness with medium to low caramel character. They have a medium body. The style may have low levels of fruity-ester flavor and aroma. Diacetyl should be absent or barely perceived. Chill haze is allowable at cold temperatures.

Original Gravity (°Plato):   1.044–1.056 (11–14 °Plato)
Apparent Extract–
Final Gravity (°Plato):   1.006–1.016 (1.5–4 °Plato)

Alcohol by weight (volume):   3.5–4.3% (4.5–5.5%)
Bitterness (IBU):   20–40
Color SRM (EBC):   11–18 (25–45 EBC)

Gold Cup Winner

Capstone ESB

Oasis Brewery and Restaurant
1095 Canyon Blvd.
Boulder, Colorado, USA 80302
Brewmaster: Bill Sherwood
Established 1992
Production: 9,000 bbl. (10,500 hl.) combined total of brew-
pub and brewery annex

Oasis also won the Oatmeal Stout category. See that category for a description of the brewery.

CHARACTER DESCRIPTION OF GOLD CUP–WINNING
CAPSTONE ESB

George Hanna

Medium, light orange-amber-hued ale. An immediate impression of American Centennial or
Cascade hops dominates the aroma, followed by
the sweetness of caramel and a subtle ale fruitiness. A pleasing blend of malt and hops balances
the aroma of this classic American-style ale. Flavor impression resounds with hop character immediately followed by
a caramel maltiness absent of toffee and toasted character.
The character suggests the use of a light-colored crystal
malt. As the first impression fades, once again the hop flavor returns, accompanied by a bitterness that remains
briefly and balances the delightful malt character. The flavor
balance evolves slowly to a refreshing yet dissipating hop
flavor. Interestingly, the hop bitterness is assertive, but it
does not overpower the overall balance of the beer. Accompanied by a medium body, the flavors dance from one to

another: hops, malt, sweet, bitter, caramel. One could use this recipe to make a variety of English ales by manipulating malt dosage and variety of hops. But I digress, for whenever the ale gets good, the good get digressing.

- **Recipe for 5 U.S. *gallons* (19 *liters*) Capstone ESB**
Targets:
Original Gravity: 1.056 (28)
Final Gravity: 1.010 (2.5)
Alcohol by volume: 6.25%
Color: 12 SRM (24 EBC)
Bittering Units: 50

ALL-GRAIN RECIPE AND PROCEDURE

8¾ lbs.	(4 kg.)	American 2-row Klages pale malt
1¼ lbs.	(0.57 kg.)	American caramel malt—30/40 Lovibond
½ lb.	(0.23 kg.)	American Cara-Pils malt
10½ lbs.	**(4.8 kg.)**	**Total grains**
8 HBU	(227 MBU)	Cascade hops (pellets)—75 minutes (bittering)
3 HBU	(85 MBU)	Cascade hops (pellets)—30 minutes (flavor)
3 HBU	(85 MBU)	Cascade hops (pellets)—20 minutes (flavor)
½ oz.	(14 g.)	Cascade hops (pellets)—10 minutes (aroma)
¼ tsp.		Irish moss
¾ c.		corn sugar for priming in bottles. Use ⅓ cup corn sugar if priming a keg.

Wyeast 1056 American Ale yeast

A single-step infusion mash is employed to mash the grains. Add 10.5 quarts (10 l.) of 167-degree F (75 C) water to the crushed grain, stir, stabilize and hold the temperature at 150 degrees F (65.5 C) for 60 minutes.

After conversion, raise temperature to 167 degrees F (75 C), lauter and sparge with 5 gallons (19 l.) of 170-degree F (77 C) water. Collect about 6 gallons (21 l.) of runoff, add bittering hops and bring to a full and vigorous boil.

The total boil time will be 75 minutes. When 30 minutes remain, add first dose of flavor hops. When 20 minutes remain, add second dose of flavor hops. When 10 minutes remain, add aroma hops and Irish moss. After a total wort boil of 75 minutes (reducing the wort volume to just over 5 gallons), turn off the heat, then separate or strain out and sparge hops. Chill the wort to 70 degrees F (21 C) and direct into a sanitized fermenter. Aerate the cooled wort well. Add a good dose of a healthy and active yeast culture for maximum attenuation potential. Ferment for 4 to 6 days in the primary. Then transfer into a secondary fermenter, chill to 55 to 60 degrees F (13–15.5 C) if possible and age for two to three weeks.

When secondary aging is complete, prime with sugar, bottle or keg. Let condition at temperatures above 60 degrees F (15.5 C) until clear and carbonated.

MALT-EXTRACT RECIPE AND PROCEDURE FOR CAPSTONE ESB

5¾ lbs. (2.6 kg.)   English extralight dried malt extract
1¼ lbs. (0.57 kg.)   American caramel malt—30/40 Lovibond
1¼ lbs. (0.57 kg.) Total grains

10   HBU (284 MBU) Cascade hops (pellets)—60 minutes (bittering)
4.5 HBU (128 MBU) Cascade hops (pellets)—30 minutes (flavor)
4   HBU (113 MBU) Cascade hops (pellets)—20 minutes (flavor)
½   oz.   (14 g.) Cascade hops (pellets)—10 minutes (aroma)

¼   tsp.   Irish moss

¾   c.   corn sugar for priming in bottles. Use ⅓ cup corn sugar if priming a keg.

Wyeast 1056 American Ale yeast

Steep crushed specialty grains in 1½ gallons (5.7 l.) water at 150 degrees F (65.5 C) for 30 minutes. Strain and sparge with enough 170-degree F (76.5 C) water to finish with a little over 2½ gallons (9.5 l.) specialty grain liquor. Add the dried malt extract and bittering hops and bring to a full and vigorous boil.

The total boil time will be 60 minutes. When 30 minutes remain, add first dose of flavor hops. When 20 minutes remain, add second dose of flavor hops. When 10 minutes remain, add aroma hops and Irish moss. After a total wort boil of 60 minutes, turn off the heat, separate or strain out and sparge hops, and direct the hot wort into a sanitized fermenter to which 2 gallons (7.6 l.) of cold water have been added. If necessary, add additional cold water to achieve a 5-gallon (19–l.) batch size. Chill the wort to 70 degrees F (21 C). Aerate the cooled wort well. Add a good dose of a healthy and active yeast culture for maximum attenuation. Ferment for 4 to 6 days in the primary. Then transfer into a secondary fermenter, chill to 55 to 60 degrees F (13–15.5 C) if possible and age for two to three weeks.

When secondary aging is complete, prime with sugar, bottle or keg. Let condition at temperatures above 60 degrees F (15.5 C) until clear and carbonated.

Silver Cup Winner

Devils Head Red
Columbine Mill Brewery
Littleton, Colorado, USA

Deep amber ale with orange hues. Sweet and floral hop aroma combine with caramel and toasted malts. Excellent toffeelike character emerges, though not intensely. Aroma also has a slight sweet cocoa character, anticipating a chewy encounter. Medium-bodied mouth feel is wholly supported with full malt flavors: caramel, cocoa and toasted malts. These gentle malt characters are neither confrontational nor acquiescent. Bitterness suggests a roast-malt-and-hop combination that would otherwise be very sharp and excessive if it weren't for the artful blend of malts used to maintain a balanced celebration. Clean overall character and memory, with the pleasant recollection of floral hops at the beginning of each sip.

Brewery formulation uses American Victory and Cara-Pils, English Munich and Belgian crystal malts combined with Nugget and Cascade hops for bitterness and aroma.

Estimated profile based on tasting
Original Gravity: 1.050 (12.5) indicated by the brewery
Final Gravity: 1.010 (2.5) indicated by the brewery
Alcohol by volume: 5% indicated by the brewery
Color: 14–15 SRM (28–30 EBC)
Bittering Units: 35–39 (30 indicated by the brewery)

Bronze Cup Winner

MacTarnahan's Ale
Portland Brewing Co.
Portland, Oregon, USA

Amber color with an orange hue. Fruity aroma is strongly suggestive of a true fresh ale. Woody, cedarlike aroma combines with a floral character, both of which are mysterious. Medium-bodied mouth feel. Subtle caramel malt contributes to overall balance. Great soft hop bitterness balances the sweetness. Floral fruitiness persists from aroma into flavor and continues to perplex. What is it? One may have to enjoy several bottles before having a clue, but then after that, what good would it be to know? This beer is for enjoying. It's not a beer that blasts your palate. Aftertaste is a lingering, playful bitterness with constant reminders of the unique fruitiness.

Brewery formulation uses crystal malts and Cascade hops for bitterness.

Estimated profile based on tasting
Original Gravity: 1.052 (13) indicated by the brewery
Final Gravity: 1.014 (3.5) indicated by the brewery
Alcohol by volume: 5% indicated by the brewery
Color: 10–12 SRM (20–24 EBC) (14 indicated by the brewery)
Bittering Units: 29–33 (30 indicated by the brewery)

CATEGORY 20:
GOLDEN ALE/CANADIAN-STYLE ALE

Golden Ales and Canadian-style Ales are a straw-colored to golden blond variation of the classic American-style pale ale. However, Golden Ale more closely approximates a lager in its crisp, dry palate, low but noticeable hop floral aroma and light body. Perceived bitterness is low to medium. Fruity esters may be perceived but do not predominate. Chill haze should be absent.

Original Gravity (°Plato):	1.045–1.056 (11–14 °Plato)
Apparent Extract– Final Gravity (°Plato):	1.008–1.016 (2–4 °Plato)
Alcohol by weight (volume):	3.2–4% (4–5%)
Bitterness (IBU):	15–30
Color SRM (EBC):	3–10 (7–20 EBC)

Gold Cup Winner

Griffon Extra Pale Ale
McAuslan Brewing Company/Brasserie McAuslan
4850, rue St. Ambroise
Bureau 100
Montréal, Québec, Canada H4A 3N8
Brewmaster: Ellen Bounsall
Established 1988
Production: 30,000 bbl. (35,000 hl.)

In the beginning there was homebrew. And just like yours, Peter McAuslan's homebrew was inspiring. Twenty-five years of quality-oriented homebrewing led Peter and his wife, Ellen Bounsall, to establish one of Montréal's foremost microbreweries.

Peter left his university registrar job to build the brewery

Ellen Bounsall, Brewmaster

Peter McAuslan

that now makes his beer. It was a partnership that was meant to be. No, not just the beer and Peter, but Peter and Ellen, who embraced the challenge of becoming head brewer for Brasserie McAuslan. Ellen, also a university senior administrator, left the world of office paperwork and politics to become one of North America's few female master brewers.

The theme of their beginning? Peter McAuslan believed (just as you do in your hometown homebrewery) that he could brew a better beer and that Quebecois consumers would provide a loyal following. A year was spent gathering funding and acquiring expertise and equipment. They gathered a team of skilled people with the temperament needed to take on the challenge of developing a new microbrewery. Alan Pugsley, the well-known British brewmaster, was hired to train the staff and develop their first beers. Soon the beers of McAuslan were receiving critical acclaim from the likes of Michael Jackson and judges in international competitions.

Beers produced by McAuslan Brewery include St. Ambroise Pale Ale, St. Ambroise Oatmeal Stout, Griffon Brown Ale and Frontenac Extra Pale Ale, as well as the award-winning Griffon Extra Pale Ale.

CHARACTER DESCRIPTION OF GOLD CUP–WINNING GRIFFON EXTRA PALE ALE

Medium golden hue with no hint of orange color. Notable head retention and foam "cling" quality. A soft, almost vanillalike hop aroma with the suggestion of an herbal-minty and floral character. American hop character, yet not citruslike. A medium body and fullness of flavor are apparent in the first impression. Flavor is wonderfully refreshing, not assaulting. Good clean malt character. A very light touch of caramel, but barely detectable. Perhaps a hint of Vienna and/or Munich malts, but more likely wheat malt—but most likely neither. Signature fruitiness from a true ale fermentation. The softness of malt and hops retrieves the alelike fruitiness in a way to bring balance to the overall flavor. Artfully achieved bitterness provides great drinkability by underpinning the immediate and final impression without assertiveness or sharpness.

- **Recipe for 5 U.S. gallons (19 liters) Griffon Extra Pale Ale**
  Targets:
  Original Gravity: 1.048 (12)
  Final Gravity: 1.012 (3)
  Alcohol by volume: 4.6%
  Color: 7 SRM (14 EBC)
  Bittering Units: 28

ALL-GRAIN RECIPE AND PROCEDURE

8	lbs.	(3.7 kg.)	Canadian 2-row Harrington pale malt
½	lb.	(.23 kg.)	wheat malt
0.4	lb.	(182 g.)	American caramel malt—40 Lovibond
8.9	**lbs.**	**(4 kg.)**	**Total grains**

4 HBU (113 MBU) American Mt. Hood hops (pellets)—60 minutes (bittering)

1.5 HBU (43 MBU) American Tettnanger hops (pellets)—30 minutes (flavor)

3 HBU (85 MBU) American Tettnanger hops (whole)—20 minutes (flavor)

3 HBU (85 MBU) American Tettnanger hops (pellets)—10 minutes (aroma)

¼ tsp. Irish moss

¾ c. corn sugar for priming in bottles. Use ⅓ cup corn sugar if priming a keg.

Wyeast 1318 London Ale III or 1968 London ESB yeast is suggested. Use a yeast that lends malty, very low sulfur, soft palate with some sweetness and low but evident fruitiness.

A single-step infusion mash is employed to mash the grains. Add 9 quarts (8.6 l.) of 170-degree F (77 C) water to the crushed grain, stir, stabilize and hold the temperature at 153 degrees F (67 C) for 60 minutes.

After conversion, raise temperature to 167 degrees F (75 C), lauter and sparge with 4.5 gallons (17 l.) of 170-degree F (77 C) water. Collect about 5.5 gallons (21 l.) of runoff, add bittering hops and bring to a full and vigorous boil.

The total boil time will be 60 minutes. When 30 minutes remain, add first dose of flavor hops. When 20 minutes remain, add second dose of flavor hops. When 10 minutes remain, add aroma hops and Irish moss. After a total wort boil of 60 minutes (reducing the wort volume to just over 5 gallons), turn off the heat, then separate or strain out and sparge hops. Chill the wort to 65 to 70 degrees F (18–21 C) and direct into a sanitized fermenter. Aerate the cooled wort well. Add an active yeast culture and ferment for 4 to 6 days in the primary. Then transfer into a secondary fermenter, chill to 55 to 60 degrees F (13–15.5 C) if possible.

When secondary aging is complete, prime with sugar, bottle or keg. Let condition at temperatures above 60 degrees F (15.5 C) until clear and carbonated.

MALT-EXTRACT RECIPE AND PROCEDURE FOR GRIFFON EXTRA PALE ALE

5¼ lbs.	(2.4 kg.)	English extralight dried malt extract
½ lb.	(0.23 kg.)	American caramel malt—40 Lovibond
0.5 lb.	**(.23 kg.)**	**Total grains**

5.5 HBU	(156 MBU)	American Mt. Hood hops (pellets)—60 minutes (bittering)
1.5 HBU	(43 MBU)	American Tettnanger hops (pellets)—30 minutes (flavor)

3   HBU   (85 MBU) American Tettnanger hops (whole)—
          20 minutes (flavor)
3   HBU   (85 MBU) American Tettnanger hops (pel-
          lets)—10 minutes (aroma)

¼   tsp.   Irish moss
¾   c.   corn sugar for priming in bottles. Use ⅓ cup
             corn sugar if priming a keg.

Wyeast 1318 London Ale III or 1968 London ESB yeast is
suggested. Use a yeast that lends malty, very low sulfur,
soft palate with some sweetness and low but evident
fruitiness.

Steep crushed specialty grains in 1.5 gallons (5.7 l.) water
at 150 degrees F (65.5 C) for 30 minutes. Strain and sparge
with enough 170-degree F (76.5 C) water to finish with a
little over 2.5 gallons (9.5 l.) specialty grain liquor. Add the
dried malt extract and bittering hops and bring to a full and
vigorous boil.

The total boil time will be 60 minutes. When 30 minutes
remain, add the first dose of flavor hops. When 20 minutes
remain, add the second dose of flavor hops. When 10 minutes remain, add aroma hops and Irish moss. After a total
wort boil of 60 minutes (reducing the wort volume to just
over 5 gallons), turn off the heat, separate or strain out and
sparge hops, and direct the hot wort into a sanitized fermenter to which 2 gallons (7.6 l.) of cold water have been
added. If necessary, add additional cold water to achieve a
5-gallon (19–l.) batch size. Chill the wort to 65 to 70 degrees
F (18–21 C). Aerate the cooled wort well. Add an active yeast
culture and ferment for 4 to 6 days in the primary. Then
transfer into a secondary fermenter, chill to 55 to 60 degrees
F (13–15.5 C) if possible and add aroma-dry hops. Allow to
age for two weeks or more.

When secondary aging is complete, prime with sugar, bottle or keg. Let condition at temperatures above 60 degrees
F (15.5 C) until clear and carbonated.

Silver Cup Winner

Molson Golden
Molson Breweries—MCI
Etobicoke, Ontario, Canada

Pale golden straw color. Aromatic sweet-malt character is so light and neutral it almost is not detectable. Light-bodied mouth feel leads directly to a clean, dry finish with only a hint of bitterness. A simple lager with finesse. So clean and simple there is not much to write about. No hop or malt character to discuss. Refreshing drinkability for a parched and thirsty palate.

Estimated profile based on tasting
Color: 4–5.5 SRM (8–11 EBC)
Bittering Units: 15–18

Bronze Cup Winner

Independence Gold
Independence Brewing Co.
Philadelphia, Pennsylvania, USA

Light amber color. A geranium-type of floral fruitiness initiates aromatic impressions. This is followed with suggestions of slightly toasted malt and an elevated alcohol content. A sharp bitterness predominates in overall balance. Clean, complex fruitiness is mild in flavor, but evident. Medium- to light-bodied mouth feel. Plenty of hop character for a golden ale. As in the aromatic impression, the aftertaste is somewhat geranium or roselike with accompanying bitterness. Crisp and clean memory, hoppy with floral hops that emphasize a delicate touch of secondary sweet maltiness.

Estimated profile based on tasting
Original Gravity: 1.049 (12.3) indicated by the brewery
Final Gravity: 1.010 (2.5) indicated by the brewery
Alcohol by volume: 4.7% indicated by the brewery
Color: 7–8 SRM (14–16 EBC) (7 indicated by the brewery)
Bittering Units: 32–36 (23 indicated by the brewery)

CATEGORY 21:
AMERICAN-STYLE BROWN ALE

American Brown Ales look like their English counterparts
but have an evident hop aroma and increased bitterness.
They have medium body. Estery and fruity-ester characters
should be subdued; diacetyl should not be perceived. Chill
haze is allowable at cold temperatures.

Original Gravity (°Plato):	1.040–1.055 (10–14 °Plato)
Apparent Extract–	
Final Gravity (°Plato):	1.010–1.018 (2.5–4.5 °Plato)
Alcohol by weight (volume):	3.3–4.7% (4–5.9%)
Bitterness (IBU):	25–60
Color SRM (EBC):	15–22 (35–90 EBC)

Gold Cup Winner:

Slow Down Brown Ale
Il Vicino Wood Oven Pizza and Brewery
3403 Central NE
Albuquerque, NM USA 87106
Brewmaster: Tom Hennessy
Brewers: Brady McKeown and Dustin Maas
Established 1993
Production: 550 bbl. (640 hl.)

Il Vicino also won the India Pale Ale category. Please see that category for a description of the brewery.

Thomas J. Hennessy

Brady McKeown, Brewmaster, and Dustin Maas, Assistant

CHARACTER DESCRIPTION OF GOLD CUP–WINNING SLOW DOWN BROWN ALE

Brown beer with a red hue. A sweeping sense of sweet caramel malt combines with a dry-roasted malt aroma along with suggestions of nuttiness. No distinct hop aroma nor ale fruitiness emerges. First flavor impression is of a perfectly balanced and targeted medium-bodied brown ale. No flaws, but plenty of suggestive character in all the right places. Roast malt, with a small hint of nutty-chocolate character. Several complex characters highlight Slow Down Brown Ale, all coming together as a singularly great expression. The low-profiled bitterness is a bit on the harsh side but does not assault or offend the palate, while still providing the punch necessary for this style of brown ale. The palate pleasantly recalls the softer characters of roast malt. Black burnt-malt character is absent. With some attention one notes a gentle floral hop flavor, which contributes to the overall pleasant character of this ale. One really needs to pay attention to identify each component, but the effort is worth it. Skillfully accomplished American-style Brown Ale.

- **Recipe for 5 U.S. gallons (19 liters) Slow Down Brown Ale**
  Targets:
  Original Gravity: 1.056 (14)

Final Gravity: 1.014 (3.5)
Alcohol by volume: 5.5%
Color: 18 SRM (36 EBC)
Bittering Units: 27

ALL-GRAIN RECIPE AND PROCEDURE
9   lbs.   (4.1 kg.)   American 2-row pale malt
½   lb.   (0.23 kg.)   American caramel malt—80 Lovibond
½   lb   (0.23 kg.)   American Cara-Pils malt
0.15   lb.   (68 g.)   American chocolate malt
0.1   lb.   (45 g.)   American roasted barley
10¼ lbs. (4.7 kg.)   Total grains

3.5   HBU   (99 MBU) German Northern Brewer hops (pellets)—75 minutes (bittering)
4   HBU   (113 MBU) German Northern Brewer hops (pellets)—30 minutes (flavor)
2.5   HBU   (71 MBU) American Cascade hops (pellets)—10 minutes (flavor)
¼   oz.   (7 g.) American Cascade hops (pellets)—dry-hop for 2 weeks (aroma)

¼   tsp.   Irish moss
¾   c.   corn sugar for priming in bottles. Use ⅓ cup corn sugar if priming a keg.
Wyeast 1187 Ringwood Ale yeast or Wyeast 1742 Swedish Ale yeast

A single-step infusion mash is employed to mash the grains. Add 10 quarts (9.5 l.) of 172-degree F (78 C) water to the crushed grain, stir, stabilize and hold the temperature at 155 degrees F (68 C) for 60 minutes.

After conversion, raise temperature to 167 degrees F (75 C), lauter and sparge with 4.5 gallons (17 l.) of 170-degree F (77 C) water. Collect about 6 gallons (23 l.) of runoff, add bittering hops and bring to a full and vigorous boil.

The total boil time will be 75 minutes. When 30 minutes remain, add the first dose of flavor hops. When 10 minutes remain, add the second dose of flavor hops and the Irish moss. After a total wort boil of 75 minutes (reducing the wort volume to just over 5 gallons), turn off the heat, then separate or strain out and sparge hops. Chill the wort to 65 to 70 degrees F (18–21 C) and direct into a sanitized fermenter. Aerate the cooled wort well. Add an active yeast culture and ferment for 4 to 6 days in the primary. Then transfer into a secondary fermenter, chill to 55 to 60 degrees F (13–15.5 C) and let age one week. Then add dry-aroma hops and let age two more weeks at 55 degrees F (13 C).

When secondary aging is complete, prime with sugar, bottle or keg. Let condition at temperatures above 60 degrees F (15.5 C) until clear and carbonated.

MALT-EXTRACT RECIPE AND PROCEDURE FOR SLOW DOWN BROWN ALE

6	lbs.	(2.7 kg.) English light dried malt extract
½	lb.	(0.23 kg.) American caramel malt—80 Lovibond
¼	lb.	(68 g.) American chocolate malt
0.1	lb.	(45 g.) American roasted barley
0.85 lb.		**(0.39 kg.) Total grains**

4.5 HBU (128 MBU) German Northern Brewer hops (pellets)—60 minutes (bittering)

5.5 HBU (156 MBU) German Northern Brewer hops (pellets)—30 minutes (flavor)

2.5 HBU (71 MBU) American Cascade hops (pellets)— 10 minutes (flavor)

¼ oz. (7 g.) American Cascade hops (pellets)—dry-hop for two weeks (aroma)

¼ tsp. Irish moss

¾ c. corn sugar for priming in bottles. Use ⅓ cup corn sugar if priming a keg.

Wyeast 1187 Ringwood Ale yeast or Wyeast 1742 Swedish Ale yeast

Steep crushed specialty grains in 1½ gallons (5.7 l.) water at 150 degrees F (65.5 C) for 30 minutes. Strain and sparge with enough 170-degree F (76.5 C) water to finish with a little over 2½ gallons (9.5 l.) specialty grain liquor. Add the dried malt extract and bittering hops and bring to a full and vigorous boil.

The total boil time will be 60 minutes. When 30 minutes remain, add the first dose of flavor hops. When 10 minutes remain, add the second dose of flavor hops and the Irish moss. After a total wort boil of 60 minutes (reducing the wort volume to just over 5 gallons), turn off the heat, then separate or strain out and sparge hops. Chill the wort to 65 to 70 degrees F (18–21 C) and direct into a sanitized fermenter. Aerate the cooled wort well. Add an active yeast culture and ferment for 4 to 6 days in the primary. Then transfer into a secondary fermenter, chill to 55 to 60 degrees F (13–15.5 C) and let age one week. Then add dry-aroma hops and let age two more weeks at 55 degrees F (13 C).

When secondary aging is complete, prime with sugar, bottle or keg. Let condition at temperatures above 60 degrees F (15.5 C) until clear and carbonated.

Silver Cup Winner

Saint Arnold Brown Ale
Saint Arnold Brewing Co.
Houston, Texas, USA

Red-copper-hued brown ale. Rich intense cocoa aroma with biscuit notes at a lower level. Any hop aroma is overshadowed in first impression by roast-malt character. Mouth feel is medium-bodied. Perception of hop bitterness is mild at first, but then, in combination with roast malt, rises to a more noticeable level. Substantial bitterness is possibly masked by bigness of malt character. Very clean overall impression with roast malt, cocoa character and balancing bitterness. As the beer warms there is a bit of floral sweetness from hops, but nothing remarkable.

Brewery formulation uses Belgian Munich, CaraMunich, Special B and chocolate malts with American Perle, Liberty and Cascade hops for bitterness and flavor.

Estimated profile based on tasting
Original Gravity: 1.053 (13) indicated by the brewery
Final Gravity: 1.013 (3) indicated by the brewery
Alcohol by volume: 5.3% indicated by the brewery
Color: 13–15 SRM (26–30 EBC)
Bittering Units: 27–32 (24 indicated by the brewery)

Bronze Cup Winner

Nightwatch Dark Ale
Maritime Pacific Brewing Co.
Seattle, Washington, USA

A rich, ruby red-brown ale. Complex fruity aroma includes hints of banana and berries. No hop aroma is perceived. First flavor impression is a complex combination of fruitiness with hints of banana and rich caramel toffee. Pleasant roast flavor underlies a balanced bitterness and a notable complement of caramel and toffee. Hop flavor is absent, but bitterness seduces like a love bite. Mouth feel is low- to medium-bodied. Clean and neutral aftertaste is unimposing.

Estimated profile based on tasting
Color: 16–19 SRM (32–38 EBC)
Bittering Units: 30–34

WORLD
BEER CUP

German Origin

CATEGORY 22:
GERMAN-STYLE KÖLSCH/
KÖLN-STYLE KÖLSCH

Kölsch, a German ale or alt-style beer, is warm-fermented using ale or lager yeasts and is aged at cold temperatures. Kölsch is characterized by a golden color and a slightly dry, winelike and subtly sweet palate. Caramel character should not be evident. The body is light. This beer has low hop flavor and aroma with medium bitterness. Wheat can be used in brewing this beer. Fruity esters should be minimally perceived if at all. Chill haze should be absent or minimal.

Original Gravity (°Plato):   1.042–1.046 (10.5–11.5 °Plato)
Apparent Extract–
Final Gravity (°Plato):   1.006–1.010 (1.5–2.5 °Plato)

Alcohol by weight (volume):   3.8–4.1% (4.4–5%)
Bitterness (IBU):   20–30
Color SRM (EBC):   3.5–5 (8–14 EBC)

Gold Cup Winner

Stoddard's Kölsch
Stoddard's Brewhouse and Eatery
111 S. Murphy Ave.
Sunnyvale, California, USA 94086
Brewmaster: Bob Stoddard
Head Brewer: Mike Gray
Established 1993
Production: 2,000 bbl. (2,300 hl.)

Stoddard's also won the English (Extra Special) Strong Bitter Category. Please see that category for a description of the brewery.

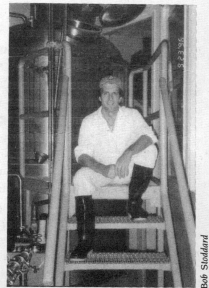

CHARACTER DESCRIPTION OF
GOLD CUP–WINNING
STODDARD'S KÖLSCH
Light golden amber in color, having notable and distinctive head retention with nice glass cling and lacy patterns of foam on the side of the glass. Subtle fruity aroma is reminiscent of Chablis/Riesling wine, perfect for this style. Absence of aggressive or notable malt character in aroma is an indication of a well-attenuated beer. Aroma

Mike Gray, Head Brewer

Bob Stoddard

is a complex combination of a minty, roselike herbal character with sweetness, though the latter is not associated with heavy malt overtones. Hop character is floral but not aggressive. Body and mouth feel are at a low to medium level. First flavor impression has a sweet impact but progresses to an extremely well balanced flavor, combining a gentle hop character and slight acidity with friendly and unassuming fruitiness. Stoddard's Kölsch resembles the character of Canadian golden ale or cream ale, but with a twist of winyness and zest.

- **Recipe for 5 U.S. gallons (19 liters) Stoddard's Kölsch**
  Targets:
  Original Gravity: 1.049 (12)
  Final Gravity: 1.010 (2.5)
  Alcohol by volume: 5%
  Color: 5 SRM (10 EBC)
  Bittering Units: 16

ALL-GRAIN RECIPE AND PROCEDURE

6	lbs.	(2.7 kg.)	American 2-row pale malt
1	lb.	(0.45 kg.)	American wheat malt
2	lbs.	(0.9 kg.)	British light Munich malt
9	**lbs.**	**(4.1 kg.)**	**Total grains**

2	HBU	(56 MBU) American Perle hops (pellets)—60 minutes (bittering)
1	HBU	(28 MBU) American Saaz hops (pellets)—60 minutes (bittering)
1	HBU	(28 MBU) American Tettnanger hops (pellets)—30 minutes (flavor)
½	oz.	(14 g.) American Hallertauer hops (pellets)—steep in finished boiled wort for 3 to 5 minutes (aroma)

¼ tsp. Irish moss
¾ c. corn sugar for priming in bottles. Use ⅓ cup corn sugar if priming a keg.

Wyeast 2565 Kölsch Ale yeast can be recommended for primary fermentation. Wyeast 2042 Danish lager yeast added during secondary lagering.

A single-step infusion mash is employed to mash the grains. Add 8 quarts (7.6 l.) of 167-degree F (75 C) water to the crushed grain, stir, stabilize and hold the temperature at 150 degrees F (65.5 C) for 60 minutes.

After conversion, raise temperature to 167 degrees F (75 C), lauter and sparge with 4 gallons (15 l.) of 170-degree F (77 C) water. Collect about 5.5 gallons (21 l.) of runoff, add bittering hops and bring to a full and vigorous boil.

The total boil time will be 60 minutes. When 30 minutes remain, add flavor hops. When 10 minutes remain, add Irish moss. After a total wort boil of 60 minutes, turn off the heat, add aroma hops and let steep 3 to 5 minutes. Then separate or strain out and sparge hops. Chill the wort to 65 to 70 degrees F (18–21 C) and direct into a sanitized fermenter. Aerate the cooled wort well. Add an active yeast culture and ferment for 4 to 7 days in the primary at temperatures no higher than 60 degrees. Then transfer into a secondary fermenter, chill to 50 to 55 degrees F (10–13 C), add a fresh and healthy culture of the lager yeast, and lager for no less than three weeks.

When secondary aging is complete, prime with sugar, bottle or keg. Let condition at temperatures above 60 degrees F (15.5 C) until clear and carbonated.

MASH-EXTRACT RECIPE AND PROCEDURE FOR STODDARD'S KÖLSCH

3¼	lbs.	(1.5 kg.)	English extralight dried malt extract
1	lb.	(0.45 kg.)	American 2-row pale malt
1	lb.	(0.45 kg.)	American wheat malt
2	lbs.	(0.9 kg.)	British light Munich malt
4	**lbs.**	**(1.9 kg.)**	**Total grains**

2.4   HBU   (66 MBU) American Perle hops (pellets)—60 minutes (bittering)

1   HBU   (28 MBU) American Saaz hops (pellets)—60 minutes (bittering)

1   HBU   (28 MBU) American Tettnanger hops (pellets)—30 minutes (flavor)

½   oz.   (14 g.) American Hallertauer hops (pellets)—steep in finished boiled wort for 3 to 5 minutes (aroma)

¼   tsp.   Irish moss

¾   c.   corn sugar for priming in bottles. Use ⅓ cup corn sugar if priming a keg.

Wyeast 2565 Kölsch Ale yeast can be recommended for primary fermentation. Wyeast 2042 Danish lager yeast added during secondary lagering.

A single-step infusion mash is employed to mash the grains. Add 4 quarts (3.8 l.) of 172-degree F (78 C) water to the crushed grain, stir, stabilize and hold the temperature at 156 degrees F (69 C) for 60 minutes.

After conversion, raise temperature to 167 degrees F (75 C), lauter and sparge with 2 gallons (7.6 l.) of 170-degree F (77 C) water. Add more water for a final volume of 3 gallons (11.5 l.) of wort. Add malt extract and bittering hops and bring to a full and vigorous boil.

The total boil time will be 60 minutes. When 30 minutes remain, add flavor hops. When 10 minutes remain, add Irish moss. After a total wort boil of 60 minutes, turn off the heat, add aroma hops and let steep for 3 to 5 minutes. Then strain out and sparge hops, and direct the hot wort into a sanitized fermenter to which 2 gallons (7.6 l.) of cold water have been added. If necessary, add additional cold water to achieve a 5-gallon (19–l.) batch size. Chill the wort to 70 degrees F (21 C). Aerate the cooled wort well. Add an active yeast culture and ferment for 4 to 7 days in the primary at temperatures no higher than 60 degrees. Then transfer into a secondary fermenter, chill to 50 to 55 degrees F (10–13 C), add a fresh and healthy culture of the lager yeast, and lager for no less than three weeks.

When secondary is complete, prime with sugar, bottle or keg. Let condition at temperatures above 60 degrees F (15.5 C) until clear and carbonated.

Silver Cup Winner

None

Bronze Cup Winner

None

CATEGORY 23:
GERMAN-STYLE BROWN ALE/
DÜSSELDORF-STYLE ALTBIER

Brown in color, this German ale may be highly hopped (though the 25–35 IBU range is more normal for the majority of Altbiers from Düsseldorf) and has a medium body and malty flavor. A variety of malts, including wheat, may be used. Hop character may be evident in the flavor. The over-all impression is clean, crisp and flavorful. Fruity esters should be low. No diacetyl or chill haze should be perceived.

Original Gravity (°Plato):   1.044–1.048 (11–12 °Plato)
Apparent Extract–
Final Gravity (°Plato):   1.008–1.014 (2–3.5 °Plato)

Alcohol by weight (volume):   3.6–4% (4.3–5%)
Bitterness (IBU):   25–48
Color SRM (EBC):   11–19 (30–45 EBC)

Gold Cup Winner

None

Silver Cup Winner

Alle Tage Alt
McNeill's Brewery Co.
Brattleboro, Vermont, USA

Brown color with red hue. Notably clear. Clean, mysterious aroma is hard to pinpoint; sweet malt character, herbal-minty-wintergreen hop character, but subtle. Velvety smooth primary caramel and secondary chocolate sweetness initially impress. Then bang, bitterness cleanly attacks and refreshes the palate. Overall impression of flavor is a clean, austere, succinct bitterness that is not cloying, astringent nor lingering. Suggestion of esters and fruitiness, but generally neutral and mild fermentation characters. Bitterness grows on you, becoming more pleasant. Medium-bodied mouth feel. Very similar to the bitter quality of the classic Düsseldorf Zum Uerige Altbier. A great example of the style, even though not awarded a gold.

This beer won the gold medal at the 1995 GABF in the German Brown Ale category.

Estimated profile based on tasting
Color: 17–20 SRM (34–40 EBC)
Bittering Units: 40–45

Bronze Cup Winner

Flagship Red Ale
Maritime Pacific Brewing Co.
Seattle, Washington, USA

Dark red-amber color with a notably rich head. Aroma is fruity with malty undertones. Light-bodied ale with a dry finish. Flavor involves a clean, moderate bitterness that tries to precede (which is unusual) the sweetness of malt. Malt sweetness character is actually quite subdued. Overall flavor is neutral, with malt taking a secondary role except for a slight refreshing and astringent roast-malt character. Caramel malt is very subdued; perhaps a lighter sweetness is perceived from a Munich malt. Hop flavor and aroma are absent.

Estimated profile based on tasting
Color: 10–13 SRM (20–26 EBC)
Bittering Units: 30–34

CATEGORY 24:
BERLINER-STYLE WEISSE (WHEAT)

This is the lightest of all the German Wheat Beers. The unique combination of a yeast-and-lactic-acid-bacteria fermentation yields a beer that is acidic, highly attenuated and very light-bodied. The carbonation of a Berliner Weisse is high and hop rates are very low. Hop character should not be perceived. Fruity esters will be evident. No diacetyl should be perceived.

Original Gravity (°Plato):   1.028–1.032 (7–8 °Plato)
Aparent Extract–
Final Gravity (°Plato):   1.004–1.006 (1–1.5 °Plato)

Alcohol by weight (volume):   2.2–2.7% (2.8–3.4%)
Bitterness (IBU):   3–6
Color SRM (EBC):   2–4 (5–10 EBC)

Gold Cup Winner

None

Silver Cup Winner

None

Bronze Cup Winner

None

CATEGORY 25:
SOUTH GERMAN–STYLE WEIZEN/ WEISSBIER

The aroma and flavor of a Weissbier are decidedly fruity and phenolic. The phenolic characteristics are often described as clove or nutmeglike, and can also be smoky or even vanillalike. These beers are made with at least 50 percent malted wheat, and hop rates are quite low. Weissbier is well attenuated and very highly carbonated, yet its relatively high starting gravity and alcohol content make it a medium- to full-bodied beer. Bananalike esters are often present. If yeast is present, the beer will appropriately have yeast flavor and a characteristically fuller mouth feel. No diacetyl should be perceived.

Original Gravity (°Plato):	1.046–1.056 (11.5–14 °Plato)
Apparent Extract– Final Gravity (°Plato):	1.008–1.016 (2–4 °Plato)
Alcohol by weight (volume):	3.9–4.4% (4.9–5.5%)
Bitterness (IBU):	10–15
Color SRM (EBC):	3–9 (8–16 EBC)

Gold Cup Winner

Sundance Hefe-Weizen
Palmer Lake Brewing Company
25 W. Cimarron
Palmer Lake, Colorado, USA 80903
Brewmaster: Kevin Cooley
Established 1994
Production: 700 bbl. (800 hl.)

Palmer Lake Brewing Company was originally established in the small mountain town of Palmer Lake by Kurt and Sallie Schoen. A desire to produce and offer for sale beers of exceptional and natural quality inspired the Schoens in this endeavor. Fashioned from a former next-door garage, this brewery at 7,323 feet won two awards in the 1996 World Beer Cup with Sundance Hefe-Weizen (Gold) and Solstice Mystery Ale (Bronze in the Belgian-Style Flanders/Oud Bruin Ale category). The Schoens have since moved their brewery to Colorado Springs. Other beers brewed by Palmer Lake Brewing Company include General Palmer's Amber Lager, Locomotive Stout, Cherry Velvet Porter and Trolley Car Ale (English bitter).

Sandi Sims

Kurt Schoen, Brewer and recipe creator, and Kevin Cooley, Brewmaster

CHARACTER DESCRIPTION OF GOLD CUP–WINNING SUNDANCE HEFE-WEIZEN

As good as wheat beer gets. Very light in color with an appropriate and appealing yeast and chill haze. Dense, rich head is excellent. Aroma expresses a distinctive balance of clove and banana with a pleasant woody, cedarlike undertone. As second and third thoughts swirl in the beer and in my head, it occurs to me that the cedarlike tone in aroma is quite a unique signature of this award-winning beer. How did they do it? Wondering brewers would like to know.

Flavors of banana, clove and acidity are expressed in equal parts; the balance is perfectly executed. Fruit and banana characters impact the palate initially, followed by the clean zestiness, and spiciness, of clove, finishing with a high-end finale. No detectable hop bitterness. Light in body and fully carbonated for style. Cedar character doesn't come through in flavor. Exceptionally clean finish adds to this beer's refreshing drinkability.

- **Recipe for 5 U.S. gallons (19 liters) Sundance Hefe-Weizen**

Targets:
Original Gravity: 1.046 (11.5)
Final Gravity: 1.008 (2)
Alcohol by volume: 5%
Color: 5 SRM (10 EBC)
Bittering Units: 9

ALL-GRAIN RECIPE AND PROCEDURE
2¾ lbs. (1.2 kg.) American 6-row pale malt
5¾ lbs. (2.6 kg.) American wheat malt
8½ lbs. (3.9 kg.) Total grains

2 HBU (56 MBU) American Hallertauer hops (pellets)—60 minutes (bittering)

¼ tsp. Irish moss
¾ c. corn sugar for priming in bottles. Use ⅓ cup corn sugar if priming a keg.
Wyeast 3068 Weihenstephan Weizen yeast

A step infusion mash is employed to mash the grains. Add 8.5 quarts (8 l.) of 136-degree F (58 C) water to the crushed grain, stir, stabilize and hold the temperature at 128 degrees F (53 C) for 30 minutes. Add 5 quarts (4.8 l.) of boiling water. Add heat to bring temperature up to 150 degrees F (65.5 C). Hold for about 60 minutes.

After conversion, raise temperature to 167 degrees F (75 C), lauter and sparge with 4 gallons (15 l.) of 170-degree F (77 C) water. Collect about 5.5 gallons (21 l.) of runoff, add bittering hops and bring to a full and vigorous boil.

The total boil time will be 60 minutes. When 10 minutes remain, add Irish moss. After a total wort boil of 60 minutes (reducing the wort volume to just over 5 gallons), turn off the heat, then separate or strain out and sparge hops. Chill the wort to 70 degrees F (21 C) and direct into a sanitized fermenter. Aerate the cooled wort well. Add an active yeast culture and ferment for 4 to 6 days in the primary. Then transfer into a secondary fermenter, chill to 60 degrees F (15.5 C). (The temperature of the fermentation and secondary aging is critical for controlling the level of banana esters and clove phenolics produced by the yeast. Every brewery must adapt a temperature schedule to its fermentation equipment. If you're not successful in producing the balance desired the first time, careful note taking will help you decide changes to make in the next batch). Allow to age for two weeks or more.

When secondary aging is complete, prime with sugar, bottle or keg. Let condition at temperatures above 60 degrees F (15.5 C) until clear and carbonated.

MALT-EXTRACT RECIPE AND PROCEDURE FOR SUNDANCE HEFE-WEIZEN

6½ lbs. (3 kg.) wheat malt extract (50% wheat/50% barley)

2.5 HBU (71 MBU) American Hallertauer hops (pellets)—60 minutes (bittering)

¼ tsp. Irish moss
¾ c. corn sugar for priming in bottles. Use ⅓ cup corn sugar if priming a keg.
Wyeast 3068 Weihenstephan Weizen yeast

Add the malt extract and bittering hops to 2½ gallons (9.5 l.) of boiling water. The total boil time will be 60 minutes. When 10 minutes remain, add Irish moss. After a total wort boil of 60 minutes, turn off the heat, separate or strain out and sparge the hops, and direct the hot wort into a sanitized fermenter to which 2 gallons (7.6 l.) of cold water have been added. If necessary, add additional cold water to achieve a 5-gallon (19–l.) batch size. Chill the wort to 70 degrees F (21 C). Aerate the cooled wort well. Add an active yeast culture and ferment for 4 to 6 days in the primary. Then transfer into a secondary fermenter, chill to 60 degrees F (15.5 C). (The temperature of the fermentation and secondary aging is critical for controlling the level of banana esters and clove phenolics produced by the yeast. Every brewery must adapt a temperature schedule to their fermentation equipment. If you're not successful in producing the balance desired the first time, careful note taking will help you decide changes to make in the next batch.) Allow to age for two weeks or more.

When secondary aging is complete, prime with sugar, bottle or keg. Let condition at temperatures above 60 degrees F (15.5 C) until clear and carbonated.

Silver Cup Winner

Edelweiss Hefetrüb
Österreichische Bräu-Aktiengesellschaft
Linz, Austria

Pale golden color with a rich, dense head and high carbonation. Aroma rejoices with a rich, fruity texture, primarily of banana. Malt and hop aroma are absent. Some of the fruitier alcohols are evident in the aroma. Mouth feel is medium-bodied. There is virtually no bitterness in flavor or aftertaste. The theme of fruitiness continues and dominates the palate along with a slight spicy clove character. These fermentive contrails characterize this beer, unlike many other styles that make their statement with the balance of hops and malt. High carbonation plays an important role in the aftertaste, refreshing the palate.

Estimated profile based on tasting
Color: 4 SRM (8 EBC)
Bittering Units: 13–16

Bronze Cup Winner

Tabernash Weiss
Tabernash Brewing Co.
Denver, Colorado, USA

Pale golden in color with an appropriate yeast haze. Good head retention. Overtures of banana and clove lightly emerge in aroma, with a generally all-around floral-fruity complexity. Very effervescent and highly carbonated. Bitterness is mild with a touch of acidity due in part to lack of hops as well as to the nature of the fermentation. Neither hop flavor nor aroma is evident. Aftertaste is yeasty. Overall balance is fruity and refreshing.

    Brewery formulation uses German wheat and Cara-Pils malts with German Northern Brewer and Perle hops for bitterness.

Estimated profile based on tasting.
Original Gravity: 1.050 (12.5) indicated by the brewery
Final Gravity: 1.011 (2.7) indicated by the brewery
Alcohol by volume: 5.7% indicated by the brewery
Color: 4–5 SRM (8–10 EBC) (6 EBC indicated by brewery)
Bittering Units: 14–16 (13 indicated by the brewery)

CATEGORY 26:
SOUTH GERMAN–STYLE DUNKEL WEIZEN/DUNKEL WEISSBIER

This beer style is characterized by a distinct sweet maltiness and a roasted-malt and chocolatelike character, but the estery and phenolic elements of a pale Weissbier still prevail. Color can range from copper-brown to dark brown. Carbonation and hop bitterness are similar to a pale South German–style

Weissbier. Usually dark barley malts are used in conjunction with dark cara or color malts, and the percentage of wheat malt is at least 50 percent. No diacetyl should be perceived.

Original Gravity (°Plato):	1.048–1.056 (12–14 °Plato)
Apparent Extract–	
Final Gravity (°Plato):	1.008–1.016 (2–4 °Plato)
Alcohol by weight (volume):	3.8–4.3% (4.8–5.4%)
Bitterness (IBU):	10–15
Color SRM (EBC):	16–23 (35–95 EBC)

Gold Cup Winner

Edelweiss Dunkel
Österreichische Bräu-Aktiengesellschaft
Alanovaplatz 5
A-2320 Schwechat, Austria
Brewmaster: Dr. Günther Seeleitner
Established 1475
Production: 342,000 bbl. (400,000 hl.)

Mag. Astrid Strange, Günter Franz

It was a time of sailing ships and sealing wax. But this is not what Johann Elesenhaimer had in mind when in 1475 he founded the brewery Hofbräu Kaltenhausen in Schwechat, just ten miles south of Salzburg in the Austrian Alps. The beer must have been good even in those days, for in 1498 the Prince Archbishop of Salzburg took over the brewery. It remained the Court Brewery of Kaltenhausen for the next 300 years. The brewery eventually fell into private hands and has survived to become, with five operating breweries, the largest producer of beer in Austria. The original brewery was rebuilt from the ground up in the midnineteenth century, but the setting remains reminiscent of medieval times. Stone buildings, narrow cobblestone streets, and the surrounding forest all contribute to the centuries-old tradition of the Österreichische Brau Aktiengeschellschaft.

Other beers produced by Österreichische Brau Aktiengeschellschaft include Edelweiss Hefetrüb, Edelweiss Kristalklar, Kaiser Pils, Kaiser Gold, Kaiser Märzen, Kaiser Premium, Kaiser Fest Bock and Kaiser Doppelmalz.

CHARACTER DESCRIPTION OF GOLD CUP–WINNING EDELWEISS DUNKEL

A light amber color with a dense, rich, creamy head and plenty of effervescence. Beer foam clinging to the sides of the glass as lace is a notable quality. This bottle-conditioned beer with yeast sediment has a healthy yeast haze. Brewed with only pale barley and wheat malt, the beer nevertheless achieves a rich amber color. The aroma of sweet cloves and fruity bananas with a touch of cinnamon spice playfully assaults the senses. Any wheat-beer lover worth his chaff would die to be chained and subjected to this dunkelweizen everlastingly. The mouth feel is medium-bodied with a dry finish. Edelweiss Dunkel first impresses as effervescent and very refreshing. Then bang! The sweet clove charges, followed by banana and then to a lesser degree by malt sweetness rounding out the entire well-balanced experience. Aftertaste briefly lingers with an herbal noble hoplike, mintlike character, but finishes very cleanly and lightly without being cloying.

I could grow to like this beer, but as with all well-made German-style Weissbiers, it would take a lot of growing. You see, I don't like this kind of beer; there had to be one, right? But don't let this stand in the way of your enjoying the best of the world's Weissbiers.

- **Recipe for 5 U.S. gallons (19 liters) Edelweiss Dunkel**
  Targets:
  Original Gravity: 1.050 (12.5)

Final Gravity: 1.010 (2.5)
Alcohol by volume: 5.2%
Color: 10 SRM (20 EBC)
Bittering Units: 13

ALL-GRAIN RECIPE AND PROCEDURE

5	lbs. (2.3 kg.)	Austrian (or available) 2-row pale malt
4	lbs. (1.8 kg.)	Austrian (or available) wheat malt
1	oz. (22 g.)	American chocolate malt (note: ounces, not pounds)
9	**lbs. (4.1 kg.)**	**Total grains**

2.5	HBU	(71 MBU) American Nugget hops (pellets)—75 minutes (bittering)
¼	oz.	(7 g.) Austrian Mühl Viertel or German Haller-tauer hops (pellets)—15 minutes (flavor)

¼	tsp.	Irish moss
¾	c.	corn sugar for priming in bottles. Use ⅓ cup corn sugar if priming a keg.

Wyeast 3068 Weihenstephan Weizen yeast can be recommended

A double-decoction mash is employed to mash the grains. If you are using alkaline or hard water, treat the water with an appropriate amount of lactic acid or refer to pages 120 to 122 of *The Home Brewer's Companion* for proceeding with a triple decoction with an acid rest.

Add 9 quarts (8.6 l.) of boiling water to the grain, raising the temperature to 122 degrees F (50 C). Then remove 4.5 quarts (4.3 l.) of the thickest mash and boil this in another vessel for 30 minutes. Stir this boiled decoction constantly. Maintain the remainder of the mash at 122 degrees F (50 C) while boiling the decocted portion. After boiling for 30 minutes, add the decoction back into the main mash vessel, raising the temperature to 150 degrees F (65.5 C). Maintain this temperature for one hour. Up to 9 quarts (8.6 l.) of boiling water may be added at any time during this period to maintain mashing temperature. After one hour remove one third of the mash (a fifty-fifty blend of thick mash and liquid) and boil about 20 minutes. Stir this second decoction constantly during the boil. When finished, return the boiled decoction to the main mash vessel, ending starch conversion by having raised the temperature to about 167 degrees F (75 C). Then lauter and sparge with 4 gallons (15 l.) of 170-degree F (77 C) water. Collect about 6 gallons (23 l.) of runoff, add bittering hops and bring to a full and vigorous boil.

The total boil time will be 90 minutes. When 15 minutes remain, add flavor hops and Irish moss. After a total wort boil of 90 minutes (reducing the wort volume to just over 5 gallons), turn off the heat, then separate or strain out and sparge hops. Chill the wort to 70 degrees F (21 C) and direct into a sanitized fermenter. Aerate the cooled wort well. Add an active yeast culture and ferment for 4 to 6 days in the primary. Then transfer into a secondary fermenter, chill to 60 degrees F (15.5 C) if possible. (The temperature of the fermentation and secondary aging is critical for controlling the level of banana esters and clove phenolics produced by the yeast. Every brewery must adapt a temperature schedule to their fermentation equipment. If you're not successful in producing the balance desired the first time, careful note taking will help you decide changes to make in the next batch.)

When secondary aging is complete, prime with sugar, bottle or keg. Let condition at temperatures above 60 degrees F (15.5 C) until clear and carbonated.

MALT-EXTRACT RECIPE AND PROCEDURE FOR EDELWEISS DUNKEL

7	lbs.	(3.2 kg.)	wheat malt extract (50% wheat, 50% barley)
1	oz.	(22 g.)	American chocolate malt (note: ounces not pounds)
1	**oz.**	**(22 g.)**	**Total grains**

3 HBU (85 MBU) American Nugget hops (pellets)—60 minutes (bittering)

¼ oz. (7 g.) Austrian Mühl Viertel or German Haller-tauer hops (pellets)—15 minutes (flavor)

¼ tsp. Irish moss

¾ c. corn sugar for priming in bottles. Use ⅓ cup corn sugar if priming a keg.

Wyeast 3068 Weihenstephan Weizen yeast can be recommended.

Steep crushed specialty grains in 1.5 gallons (5.7 l.) water at 150 degrees F (65.5 C) for 30 minutes. Strain and sparge with enough 170-degree F (76.5 C) water to finish with a little over 2.5 gallons (9.5 l.) specialty grain liquor. Add the dried malt extract and bittering hops and bring to a full and vigorous boil.

The total boil time will be 60 minutes. When 15 minutes remain, add flavor hops and Irish moss. After a total wort boil of 60 minutes, turn off the heat, separate or strain out and sparge hops, and direct the hot wort into a sanitized fermenter to which 2 gallons (7.6 l.) of cold water have been added. If necessary, add additional cold water to achieve a 5-gallon (19–l.) batch size. Chill the wort to 70 degrees F (21 C). Aerate the cooled wort well. Add an active yeast culture and ferment for 4 to 6 days in the primary. Then transfer into a secondary fermenter, chill to 60 degrees F (13–15.5 C) if possible. (The temperature of the fermentation and secondary aging is critical for controlling the level of banana esters and clove phenolics produced by the yeast. Every brewery must adapt a temperature schedule to their fermentation equipment. If you're not successful in producing the balance desired the first time, careful note taking will help you decide changes to make in the next batch.)

When secondary aging is complete, prime with sugar, bottle or keg. Let condition at temperatures above 60 degrees F (15.5 C) until clear and carbonated.

Silver Cup Winner

Paulaner Dunkel Weizen
Paulaner Brewery
Munich, Germany

Light brown with orange-amber hue. Remarkable and notably rich, dense head. From the opening of the bottle there is an indication of appropriately high carbonation. Everything you have come to expect in a dunkelweizen is evident in the aroma; an undertone of chocolate roasted malts accompanies a blessing of spice, clove and banana character. In addition, a complex fruitiness invites all Weizenbier fans. All the while the rich head continues to confront your lips and lather your mustache (and if you don't have one, you will now). The sweetness and fullness of the dark-roasted malts bring into perfect balance the sharpness of clove, banana and fruitiness with barely perceived low bitterness. Some roast-malt bitterness comes through without being astringent. Aftertaste is pleasantly and slightly acidic/sour, attributable to both lack of bitterness and fermentation character. Finally, a calming cocoa character thanks you in the aftertaste that lingers briefly with its rich malt sweetness. A classic.

Estimated profile based on tasting
Color: 12–14 SRM (24–28 EBC)
Bittering Units: 13–15

Bronze Cup Winner

None

CATEGORY 27:
SOUTH GERMAN–STYLE
WEIZENBOCK/WEISSBOCK

This style can be either pale or dark and, like a bottom-fermented bock, has a high starting gravity and alcohol content. The malty sweetness of a Weizenbock is balanced with the clovelike phenolic and fruity-estery-banana element to produce a well-rounded aroma and flavor. As is true with all German Wheat Beers, hop rates are low and carbonation is high. It has a medium to full body. If dark, a mild roast-malt character should emerge in flavor and to a lesser degree in the aroma. No diacetyl should be perceived.

Original Gravity (°Plato):	1.066–1.080 (16–20 °Plato)
Apparent Extract– Final Gravity (°Plato):	1.016–1.028 (4–7 °Plato)
Alcohol by weight (volume):	5.5–7.5% (6.9–9.3%)
Bitterness (IBU):	10–15
Color SRM (EBC):	5–30 (14–120 EBC)

Gold Cup Winner

Aventinus
Weißbierbrauerei G. Schneider & Sohn KG
Emil-Ott Straße 1-5
93309 Kelheim, Germany
Brewmaster: H. P. Drexler
Established 1872
Production: 385,000 bbl. (450,000 hl.)

At the very same moment Christopher Columbus discovered the New World, Bavarians were enjoying almost nothing but

Peggy Nelson

Schneider Brewery in Kelheim

Weissbier. Incredible as it may seem, its popularity almost drove it to extinction.

In the seventeenth century, brewing was a privilege reserved for nobility. Duke Maximilian opened the famous Hofbräuhaus in Munich, intending to increase the popularity of Weissbier by making it available to the masses. His success was copied by several other brewers, thus fueling its popularity at the time. But lager beers and fashion-conscious consumers challenged the popularity of Weissbiers throughout Bavaria. Perhaps because it was considered a common beer, many wheat-beer breweries went out of business. Weissbier almost became an extinct style of beer.

In 1872 Georg Schneider bought an existing brewery, and with it the privilege for brewing Weissbier. It wasn't too long before he revived its popularity with good business sense and an improvement in quality. Georg Schneider is credited by many as saving and helping to popularize this style of beer with quality wheat beers such as Aventinus, introduced as Bavaria's first wheat double bock (doppelbock) in 1907.

Still brewed today, the original Schneider Weisse is considered a world classic by many. Schneider Weisse, Schneider Weisse Weizenhell (paler than the original), Schneider Weisse Kristal (yeast-free) and Aventinus are now exported to over twenty countries, including the United States.

CHARACTER DESCRIPTION OF GOLD CUP–WINNING AVENTINUS

Simple light brown color with a slight amber hue. Not particularly red or orange. Bottle-conditioned with a well-settled and compacted yeast. Yeast haze is dependent on care of pour. Classic South German–style wheat-beer aroma is profoundly evident; clove and subtle banana character with some smoky and chocolate malt notes. Overall very fruity with a flavor balance of roast malts and banana and chocolate characters. An obvious prickly and numbing sensation of alcohol is felt on the palate. Thoroughly a bock-strength brew. There is no hop aroma or flavor. Bitterness is achieved with noble hops and is appropriately very low, yet adequate for balancing the refreshing undertone of acidity. Skillfully accomplished. Mouth feel and body are at a medium level, bordering on fullness. Aftertaste is a memorable banana fruitiness and tingling of alcohol while aroma continues with clove, banana and subtle smokiness. As with all Weissbiers, the quality of the yeast and controlled temperature fermentation are essential and must be executed without compromise.

- **Recipe for 5 U.S. gallons (19 liters) Aventinus**
  Targets:
  Original Gravity: 1.072 (36)
  Final Gravity: 1.014 (3.5)
  Alcohol by volume: 8%
  Color: 21 SRM (42 EBC)
  Bittering Units: 15

ALL-GRAIN RECIPE AND PROCEDURE

7¼ lbs.	(3.3 kg.)	German/Bavarian wheat malt
6 lbs.	(2.7 kg.)	German/Bavarian 2-row malt
0.3 lb.	(136 g.)	German chocolate malt—400 Lovibond or 0.1 lb. German 1,200 Lovibond roasted malt
13.55 lbs.	**(6.2 kg.)**	**Total grains**

3.7 HBU (101 MBU) German Hallertauer hops (pellets)—90 minutes (bittering)

¼   oz.   (7g.) German Hallertauer hops (pellets)—
          steep in finished boiled wort for 2 to 3 min-
          utes (aroma)

¼   tsp.   Irish moss
¾   c.   corn sugar for priming in bottles. Use ⅓ cup
          corn sugar if priming a keg.
Wyeast 3068 Weihenstephan Weizenbier yeast

A step infusion mash is employed to mash the grains.
Add 13.5 quarts (12.8 l.) of 138-degree F (58 C) water to the
crushed grain, stir, stabilize and hold the temperature at
128 degrees F (53 C) for 30 minutes. Add 7 quarts (6.7 l.)
of boiling water, add heat to bring temperature up to 152
degrees F (67 C) and hold for about 60 minutes.

After conversion, raise temperature to 167 degrees F (75
C), lauter and sparge with 4 gallons (15.2 l.) of 170-degree
F (77 C) water. Collect about 6 gallons (23 l.) of runoff, add
bittering hops and bring to a full and vigorous boil.

The total boil time will be 90 minutes. When 10 minutes
remain, add Irish moss. After a total wort boil of 90 minutes
(reducing the wort volume to just over 5 gallons), turn off
the heat, add aroma hops and let steep 2 to 5 minutes.
Then separate or strain out and sparge hops. Chill the wort
to 65 to 70 degrees F (18–21 C) and direct into a sanitized
fermenter. Aerate the cooled wort well. Add an active yeast
culture and ferment for 4 to 6 days in the primary. Then
transfer into a secondary fermenter and allow to age for two
weeks or more. (The temperature of the fermentation and
secondary aging is critical for controlling the level of banana
esters and clove phenolics produced by the yeast. Every
brewery must adapt a temperature schedule to their fermentation equipment. If you're not successful in producing the
balance desired the first time, careful note taking will help
you decide changes to make in the next batch.)

When secondary aging is complete, prime with sugar, bot-

tle or keg. Let condition at temperatures above 60 degrees F (15.5 C) until clear and carbonated.

MALT-EXTRACT RECIPE AND PROCEDURE FOR AVENTINUS

10	lbs. (4.5 kg.)	wheat malt syrup extract (50% wheat, 50% barley malt)
0.3	lb. (136 g.)	German chocolate malt—400 Lovibond or 0.1 lb. German 1,200 Lovibond roasted malt
0.3	**lb. (136 g.)**	**Total grains**

4.5	HBU (128 MBU)	German Hallertauer hops (pellets)—60 minutes (bittering)
¼	oz.	(7 g.) German Hallertauer hops (pellets)—steep in finished boiled wort for 2 to 3 minutes (aroma)

¼	tsp.	Irish moss
¾	c.	corn sugar for priming in bottles. Use ⅓ cup corn sugar if priming a keg.

Wyeast 3068 Weihenstephan Weizenbier yeast

Steep crushed specialty grains in 2 quarts (1.9 l.) water at 150 degrees F (65.5 C) for 30 minutes. Strain and sparge with enough 170-degree F (76.5 C) water to finish with a little over 2.5 gallons (9.5 l.) specialty grain liquor. Add the malt extract and bittering hops and bring to a full and vigorous boil.

The total boil time will be 60 minutes. When 10 minutes remain, add Irish moss. After a total wort boil of 60 minutes, turn off the heat, add aroma hops and let steep 2 to 5 minutes. Then separate or strain out and sparge hops. Chill the wort to 65 to 70 degrees F (18–21 C) and direct into a sanitized fermenter. Aerate the cooled wort well. Add an active yeast culture and ferment for 4 to 6 days in the primary. Then transfer into a secondary fermenter and allow to age for two weeks or more. (The temperature of the fermentation and secondary aging is critical for controlling the level of banana esters and clove phenolics produced by the yeast. Every brewery must adapt a temperature schedule to their fermentation equipment. If you're not successful in producing the balance desired the first time, careful note taking will help you decide changes to make in the next batch.)

When secondary aging is complete, prime with sugar, bottle or keg. Let condition at temperatures above 60 degrees F (15.5 C) until clear and carbonated.

Silver Cup Winner

None

Bronze Cup Winner

Wild Pitch Weizen Bock
Sandlot Brewery at Coors Field
Denver, Colorado, USA

Bubbles struggle to reach the surface, indicating a full-bodied brew. Orange-amber in color with lots of chill haze. Explosive aroma of cloves underpinned with banana aroma. Medium- to full-bodied mouth feel. Charged with full caramel malt and banana flavor. The bitterness is unusually obvious for a wheat beer but not so predominant that it interferes with the classic wheat-beer character. With so much body and malt flavor, the higher hop character is necessary to balance an otherwise potentially cloying beer. Aftertaste impression if fruity and sweet. A most definitive Weizenbock.

Estimated profile based on tasting
Color: 9–11 SRM (18–22 EBC)
Bittering Units: 18–22

Belgian and French Origin

CATEGORY 28:
BELGIAN-STYLE FLANDERS/
OUD BRUIN ALES

A light- to medium-bodied deep copper to brown ale characterized by a slight vinegarlike or lactic sourness and spiciness. A fruity-estery character is apparent with no hop flavor or aroma. Flanders Brown Ales have low to medium bitterness. Very small quantities of diacetyl are acceptable. Roasted malt character in aroma and flavor can be at low levels.

Original Gravity (°Plato):   1.044–1.056 (11–14 °Plato)

Apparent Extract–
Final Gravity (°Plato):   1.008–1.016 (2–4 °Plato)

Alcohol by weight (volume):   3.8–4.4% (4.8–5.2%)
Bitterness (IBU):   15–25
Color SRM (EBC):   12–18 (40–90 EBC)

Gold Cup Winner

Liefmans Goudenband
Brouwerij Liefmans (Liefmans Brewery)
Aalstrstraat 200
9700 Oudenaarde, Belgium
Brewmaster: Filip de Velder
Established 1679
Production: 25,600 bbl. (30,000 hl.)

The exact date the brewery was founded is not known, but an excise registry confirms that the brewery was in full activity in 1679. Over 300 years later, beer is still fermented at Liefmans in Oudenaarde. In the early 1990s Brewery Riva of Dentrergem bought Liefmans Brewery. It is at Brewery Riva

At left, Filip de Velder, Brewmaster

Annick De Splenter

that the wort is produced, but so unique are the environment and fermentation at the original brewery that the unfermented wort is trucked to Oudenaarde for a complex fermentation involving a century-old combination of yeast and bacteria. Fermentation occurs first in open tanks and secondarily in closed tanks. Young beer is stored for several months, after which blends of different batches are combined to establish the consistent character of Liefmans brown ales. Like homebrew, Liefmans Goudenband is bottle-conditioned with a small dose of sugar, then is stored in vast cellars for at least three months before being released for sale. Interestingly, the Gold Cup–winning Goudenband is a departure from the 5½ percent–alcohol Goudenband of years past, presently weighing in at a very strong 8 percent alcohol by volume. Liefmans Frambozen (raspberry) and Liefmans Kriek (cherry) are produced from a brown ale wort of lesser strength and then combined with fruit after three or four months of fermentation and aging. The addition of fruit initiates new fermentation, after which the beer is aged further before bottling. The complex flavors of all three Liefmans products develop with age. It certainly is a matter of personal preference deciding at which point to enjoy these special and unique products.

CHARACTER DESCRIPTION OF GOLD CUP–WINNING LIEFMANS
GOUDENBAND

The experience begins with the pop of a cork, then continues with greetings from a brown-hued ale with glimpses of orange. Complex and intelligible, an exotic aroma becomes a prelude to an anticipated acidic-sour flavor. But first a slight effusion of aromatic, sweet, caramel, toasted/roasted malt character glimmers through the bold acidity and fruitiness. There is a sense of mysteriousness and antiquity evoked as one dwells with the exotic. Then, like a whisper, a chocolatelike aroma emerges in harmony with the overall complex aromatic character. Any hop character is all but lost in the balance of fruitiness and malt. If one stays with the aroma long enough, the sour-acidic aroma becomes a cross between oxidized-acetic acid and lactic acid that is woody in nature; the acidity is not extreme in sharpness. Like a sweet and sour plum, the skin harboring the acidity while the flesh awaits voluptuously.

First flavor impression is acidic but then is suggestive of the malty brown-ale character it might have originally had, save for its gentle migration to acidification. Toasted and roast malts soften the character of the sourness. Sourness is punctual, and does not linger. Hop bitterness is very subdued and at threshold, thus creating a complex harmony with fermented sourness and malt sweetness. High end note of sourness without pucker.

Body is medium to low in the mouth feel. The recurring theme of sourness tends to clean the palate with a sensation of dryness.

- ## Recipe for 5 U.S. gallons (19 liters) Liefmans Goudenband

  Targets:
  Original Gravity: 1.079 (19)
  Final Gravity: 1.017 (4.2)
  Alcohol by volume: 8%
  Color: 20 SRM (40 EBC)
  Bittering Units: 15

ALL-GRAIN RECIPE AND PROCEDURE

2½ lbs. (1.1 kg.)	French 2-row Prisma pale malt
4 lbs. (1.8 kg.)	French 2-row Alexis pale malt
4 lbs. (1.8 kg.)	French Munich malt—7 Lovibond (15 EBC)

Note: 6½ lbs. (3.0 kg.) of American 2-row can be substituted for pale malt. Use other European Munich malt if French is not available.

2 lbs. (0.9 kg.)	Belgian CaraMunich—70 Lovibond (150 EBC)

2½ lbs. (1.1 kg.) Belgian CaraVienne—22 Lovibond (60 EBC)

15 lbs. (7.7 kg.) Total grains

3 HBU (85 MBU) German Hallertauer hops (pellets)— 120 minutes (bittering)

2 HBU (56 MBU) English Kent Goldings hops (pellets)—20 minutes (flavor)

¼ tsp. Irish moss

¾ c. corn sugar for priming in bottles. Use ⅓ cup corn sugar if priming a keg.

Recommend culturing yeast from a bottle of Goudenband and then combining with a healthy fresh culture of a fruity ale yeast such as Wyeast 1098. Do not use Belgian lambic yeast or Belgian or German Ale yeasts that produce banana esters or clove phenols.

A single-step infusion mash is employed to mash the grains. Add 15 quarts (14 l.) of 167-degree F (75 C) water to the crushed grain, stir, stabilize and hold the temperature at 150 degrees F (65.5 C) for 60 minutes.

After conversion, raise temperature to 167 degrees F (75 C), lauter and sparge with 5 gallons (19 l.) of 170-degree F (77 C) water. Collect 7 gallons (27 l.) of runoff, add bittering hops and bring to a full and vigorous boil.

The total boil time will be 120 minutes. When 20 minutes remain, add flavor hops. When 10 minutes remain, add Irish moss. After a total wort boil of 120 minutes (reducing the wort volume to just over 5 gallons), turn off the heat, then separate or strain out and sparge hops. Chill the wort to 70 degrees F (21 C) and direct into a sanitized fermenter. Aerate the cooled wort well. Add an active yeast culture and ferment for 4 to 6 days in the primary. Then transfer into a secondary fermenter and let age for two or three months, preferably at about 70 degrees F (21 C).

When secondary aging is complete, prime with a fresh

yeast culture and sugar, bottle or keg. Let condition at temperatures above 60 degrees F (15.5 C) until clear and carbonated. Aging will create complexity and allow character to evolve.

MASH-EXTRACT RECIPE AND PROCEDURE FOR LIEFMANS GOUDENBAND

7	lbs.	(3.2 kg.)	English amber dried malt extract
3	lbs.	(1.4 kg.)	2–row pale malt
1	lb.	(0.45 kg.)	Munich malt
½	lb.	(0.23 kg.)	Belgian CaraMunich—70 Lovibond (150 EBC)
1	lb.	(0.45 kg.)	Belgian CaraVienne—22 Lovibond (60 EBC)
5½	**lbs.**	**(2.5 kg.)**	**Total grains**

4	HBU	(113 MBU) German Hallertauer hops (pellets)—75 minutes (bittering)
2.5	HBU	(71 MBU) English Kent Goldings hops (pellets)—20 minutes (flavor)

¼	tsp.	Irish moss
¾	c.	corn sugar for priming in bottles. Use 1/3 cup corn sugar if priming a keg.

Recommend culturing yeast from a bottle of Goudenband and then combining with a healthy fresh culture of a fruity ale yeast such as Wyeast 1098. Do not use Belgian lambic yeast or Belgian or German Ale yeasts that produce banana esters or clove phenols.

A single-step infusion mash is employed to mash the grains. Add 5½ quarts (5 l.) of 167-degree F (75 C) water to the crushed grain, stir, stabilize and hold the temperature at 150 degrees F (65.5 C) for 60 minutes.

After conversion, raise temperature to 167 degrees F (75 C), lauter and sparge with 2½ gallons (7.6 l.) of 170-degree F (77 C) water. Collect about 3 gallons (11.5 l.) of runoff.

Add malt extract and bittering hops and bring to a full and vigorous boil.

The total boil time will be 75 minutes. When 20 minutes remain, add flavor hops. When 10 minutes remain, add Irish moss. After a total wort boil of 75 minutes (reducing the wort volume to just over 5 gallons), turn off the heat, then separate or strain out and sparge hops, and direct the hot wort into a sanitized fermenter to which 2 gallons (7.6 l.) of cold water have been added. If necessary, add additional cold water to achieve a 5-gallon (19-l.) batch size. Chill the wort to 70 degrees F (21 C). Aerate the cooled wort well. Add an active yeast culture and ferment for 4 to 6 days in the primary. Then transfer into a secondary fermenter and let age for two or three months, preferably at about 70 degrees F (21 C).

When secondary aging is complete, prime with a fresh yeast culture and sugar, bottle or keg. Let condition at temperatures above 60 degrees F (15.5 C) until clear and carbonated. Aging will create complexity and allow character to evolve.

Silver Cup Winner

None

Bronze Cup Winner

Solstice Mystery Ale
Palmer Lake Brewing Co.
Palmer Lake, Colorado, USA

Very dark brown color with a ruby red tint. Extremely exotic fruity and licoricelike aroma. With hints of mysterious Belgian yeast character, the aroma of Solstice Mystery Ale suggests acidity and erotic fruitiness. Medium body with a clean finish. Rich flavor includes licorice-anise character followed by a deep, dark caramel fullness. No real cocoa or roast-malt characters emerge to confuse you. The hop level is adequate for balance, but is not the primary character of this beer. Wild, wild fruitiness that is almost plumlike. A wonderful combination of medium body, licorice, sweet maltiness and balanced bitterness. Curiously, everything is both definitive and complex. There are no fermentation flaws. A very creative and refreshing beer in the strange exotic style and spirit of Belgian Oud Bruins.

Brewery formulation uses Belgian CaraVienne, CaraMunich and Special "B" malts along with corn, oats, vanilla, coriander, anise and curaçao orange peel. Four-year-old American Mt. Hood hops are used for bitterness. Yeast used is a Belgian Rodenbach culture. In the words of the brewer, "We do not try to approximate style too closely."

Estimated profile based on tasting
Original Gravity: 1.060 (15) indicated by the brewery
Final Gravity: 1.007 (2 indicated by the brewery)
Alcohol by volume: 7% indicated by the brewery
Color: 26–29 SRM (52–58 EBC)
Bittering Units: 30–33 (30 indicated by the brewery)

CATEGORY 29:
BELGIAN-STYLE ABBEY ALE

This medium- to full-bodied, dark amber to brown ale has a malty sweetness and nutty, chocolate, roast-malt aroma. A faint hop aroma is acceptable. Medium to full body. Dubbels are also characterized by low bitterness and no hop flavor. Very small quantities of diacetyl are acceptable. Fruity esters (especially banana) are appropriate at low levels. Head is dense and mousselike.

Original Gravity (°Plato):   1.050–1.070 (12.5–17.5 °Plato)
Apparent Extract–
Final Gravity (°Plato):   1.012–1.016 (3–4 °Plato)
Alcohol by weight (volume):   4.8–6.0% (6.0–7.5%)
Bitterness (IBU):   18–25
Color SRM (EBC):   10–14 (25–40 EBC)

Gold Cup Winner

None

Silver Cup Winner

Abbey Belgian Style Ale
New Belgium Brewing Co.
Ft. Collins, Colorado, USA

Bottle-conditioned, dark amber-brown ale with a deep red tint. Effusively complex fruitiness includes bananas. Hop aroma is absent. Mouth feel is medium-bodied. Fruitiness continues in the flavor. The perception of bitterness is relatively low, but with high maltiness and fruitiness, this impression may be deceiving because of the malt sweetness and banana. Many interesting flavors engage the palate in several stages. Fruitiness and banana are main characters, with fullness of pale, caramel and brown malts playing a supporting role. Bitterness is always in the background, emerging only for a cameo role and sensible lingering in the aftertaste. Surely the bottle refermentation is key to evolving the complex character of this wonderful ale. Worthy of the honor it has received.

Estimated profile based on tasting
Color: 19–21 SRM (38–42 EBC)
Bittering Units: 28–30

Bronze Cup Winner

St. Bernardus Tripel
Brouerij St. Bernardus
Watou, Belgium

Yeast sediment is firm, enabling a clear pour. Wonderful thick head develops, with Belgian lace forming on the sides of the glass as the foam retreats to a mustache level. Light golden in color. Intense, wildly pleasing and evocative fruitiness. But not a wild fruitiness. There is a difference. Banana is evident in aroma, but other well-aged, bottle-conditioned characters emerge to create a complex fruity character, joined eventually by a soothing mild vanilla character and a tangential flare of esterlike alcohols. Hop aroma is not apparent, but possibly there, overshadowed by complex fruitiness. Full-bodied mouth feel is accompanied by malt sweetness and secondary fruitiness in flavor. Bitterness is slow to develop, creeping upon the palate as an afterthought, joining the overall impression to help balance the intensity of alcohol, maltiness and fruitiness. Some higher alcohols create a slight solventlike impact. Full-flavored, full-bodied and very complex.

Brewery formulation uses Belgian dark malts with Belgian Northern Brewer and Belgian Saaz for bitterness, flavor and aroma.

Estimated profile based on tasting
Original Gravity: 1.075 (18) indicated by the brewery
Final Gravity: 1.026 (6.6) indicated by the brewery
Alochol by volume: 7.5% indicated by the brewery
Color: 5–6 SRM (10–12 EBC) (27 EBC indicated by the brewery)
Bittering Units: 35–39 (30 indicated by the brewery)

CATEGORY 30:
BELGIAN-STYLE PALE ALE

Characterized by low but noticeable hop bitterness, flavor and aroma. Light to medium body and low malt aroma. Golden to deep amber in color. Noble hop types commonly used. Low to medium fruity esters evident in aroma and flavor. Low caramel or toasted-malt flavor acceptable. No diacetyl should be perceived. Chill haze is allowable at cold temperatures.

Original Gravity (°Plato):	1.044–1.054 (11–13.5 °Plato)
Apparent Extract–Final Gravity (°Plato):	1.008–1.014 (2–3.5 °Plato)
Alcohol by weight (volume):	3.2-5.0% (4.0–6.0%)
Bitterness (IBU):	20–30
Color SRM (EBC):	3.5–12 (8–30 EBC)

Gold Cup Winner

None

Silver Cup Winner

Fat Tire Amber Ale
New Belgium Brewing Co.
Ft. Collins, Colorado, USA

Clear and alluring red-copper-amber color. Notable and intense biscuitlike sweetness followed by caramel malt character in the aroma. Biscuit aroma masks all evidence of hop aroma. Medium-bodied mouth feel. Flavor is very malty with a biscuitlike and caramel follow-up as in aroma. What is really remarkable is that I've had this beer so many times before on tap at local restaurants and bars under conditions that do no justice to the true character of this beer. It is served too cold. Only at temperatures of 45 degrees F (7 C) or above can Fat Tire's true character really emerge. At colder temperatures malt character is numbingly absent. Fruitiness is suggested only if actively sought. Without being overdone, bitterness complements the malt richness. It is used to balance, not as a statement for bitterness. Clean aftertaste, finishes with biscuitlike and caramel memory. Medium body.

Estimated profile based on tasting
Color: 10–12 SRM (20–22 EBC)
Bittering Units: 30–33

Bronze Cup Winner

Orval Trappist Ale
Orval Trappist Monastery
Florenville, Belgium

Amber gold color. Refermented bottle-conditioned ale. A notable pour forms a rich, dense head that lingers and clings to the side of the glass. Aroma is intensely fruity, with a secondary role performed by a rich charge of hops. Spicy floral, yet subdued coriander aroma. Well-attenuated fermentation creates a knowingly deceiving very light body. Winy fruitiness. Hop flavor is recognized, but could easily be neglected for all the complexity of fruitiness and the special combination of brewery yeasts and bacteria. High in alcohol, yet well balanced. Hints of wild character, though not out of control. Aftertaste is a pleasant bitterness followed by a visit from a wild cherry, plumlike fruitiness, other complex esters and a floral hop and coriander flavor. And then hop and coriander are recalled.

Brewery formulation uses French CaraVienne malt with Hallertauer and Styrian Goldings for bitterness, flavor and aroma.

Estimated profile based on tasting
Original Gravity: 1.055 (13.5) indicated by the brewery
Final Gravity: 1.027 (0.7) two months after bottling indicated by the brewery
Alcohol by volume: 6.2% indicated by the brewery
Color: 9–11 SRM (18–22 EBC) (22 EBC indicated by the brewery)
Bittering Units: 35–40 (32 indicated by the brewery)

CATEGORY 31:
BELGIAN-STYLE STRONG ALE

Belgian Strong Ales are often vinous, with darker styles typically colored with dark candy sugar. The perception of hop bitterness can vary from low to high, while hop aroma and flavor are very low. These beers are highly attenuated and have a highly alcoholic character, being medium-bodied rather than full-bodied. Very low or no diacetyl is perceived. Chill haze is allowable at cold temperatures.

Original Gravity (°Plato):	1.064–1.096 (16–24 °Plato)
Apparent Extract— Final Gravity (°Plato):	1.012–1.024 (3–6 °Plato)
Alcohol by weight (volume):	5.6–8.8% (7.0–11.0%)
Bitterness (IBU):	20–50
Color SRM (EBC):	3.5–20 (8–80 EBC)

Gold Cup Winner

Pauwel Kwak
Brouwerij Bosteels
Kerkstraat 92
B-9255, Buggenhout, Belgium
Brewmaster: Antoine Bosteels
Established 1791
Production: 26,000 bbl. (30,000 hl.)

Award-winning Pauwel Kwak may be the quintessential winner of the World Beer Cup as far as any homebrewer is concerned. The beer and brewery were established over 200 years ago by the Flemish master brewer Pauwel Kwak. Quirky and unique, the Kwak ale, still true to its original recipe, was formulated as a homebrew in every sense of our tradition. Brewed to refresh travelers and stagecoach drivers, Pauwel Kwak was a welcome brew for all who passed through De Hoorn ("The Horn"), an infamous inn located near his stable and stagecoach stop. It was homebrewed in small batches with unique characters only homebrewers could fully appreciate. "Relax, don't worry and have a homebrew" surely would have been uttered by horsemen, coachmen, postilions (*side-kicks* to us Americans) and weary travelers of the Flemish roads of old.

Interestingly, the brew became known—and is still sometimes referred to today—as the "coachmen's brew." So important were his coachmen and postilion customers that Pauwel Kwak specially designed a drinking vessel for them. It is a cone-shaped glass with a spherical bottom that resembles a small yard of ale. This glass could easily be passed up to the masters of the road without requiring them to dismount their carriages, thus circumventing the Napoleonic Code, which stated that coachmen could not get off to join their passengers in drink.

After Pauwel Kwak's retirement from brewing, ownership passed into the hands of the Bosteels family. They have been continuously operating the brewery for six family generations. Over 80 percent of the brewery's production is the fabled Pauwel Kwak. Only after World War II was Pauwel Kwak "exported" outside the village of Buggenhout, allowing people in select locations around the world to enjoy this memorable product. Fully two thirds of the brewery's production is now exported, some of which may find its way to the United States by the time this book is published.

CHARACTER DESCRIPTION OF GOLD CUP–WINNING PAUWEL KWAK

Orange–light amber color with no red hue. Very rich, dense, white foam with "clingability" to the sides of the glass. Aroma is complex with a light blend of banana, fruitiness, chocolate and caramel. First impression of flavor is full-bodied with a fruity and complex character. Malty, caramel and slight chocolate notes are evident. The warmth of alcohol, accurately stated on the label at 8 percent by volume, certainly is perceived in tasting. There is no hop aroma or flavor to speak of, though bitterness is evident, balancing the overall character. The gentle, soft Goldings nature of hops and bitterness is not a primary flavor component.

- ### *Recipe for 5 U.S. gallons (19 liters) Pauwel Kwak*
  Targets:
  Original Gravity: 1.079 (19)
  Final Gravity: 1.014 (3.5)
  Alcohol by volume: 8%
  Color: 14 SRM (28 EBC)
  Bittering Units: 15

ALL-GRAIN RECIPE AND PROCEDURE

5	lbs. (2.3 kg.)	French Pilsener malt or other Pilsener malt
10	lbs. (4.5 kg.)	Belgian Munich malt—7 Lovibond (15 EBC)
15	**lbs. (6.8 kg.)**	**Total grains**

3   HBU   (85 MBU) English Challenger hops (pellets)—
             120 minutes (bittering)
2.5   HBU   (71 MBU) European Styrian Goldings hops
             (pellets)—15 minutes (flavor/aroma)
1   HBU   Czech Saaz hops (pellets)—15 minutes (flavor/
             aroma)

¼   tsp.   Irish moss
¾   c.   corn sugar for priming in bottles. Use ⅓ cup
             corn sugar if priming a keg.
Wyeast 1214 Belgian Ale yeast suggested; it is high-gravi-
ty- and ester-producing.

A single-step infusion mash is employed to mash the
grains. Add 15 quarts (14 l.) of 167-degree F (75 C) water
to the crushed grain, stir, stabilize and hold the temperature
at 150 degrees F (65.5 C) for 60 minutes.

After conversion, raise temperature to 167 degrees F
(75 C), lauter and sparge with 5 gallons (19 l.) of 170-degree
F (77 C) water. Collect 7 gallons (27 l.) of runoff, add bittering hops and bring to a full and vigorous boil.

The total boil time will be 120 minutes. When 20 minutes
remain, add flavor hops. When 10 minutes remain, add Irish
moss. After a total wort boil of 120 minutes (reducing the
wort volume to just over 5 gallons), turn off the heat, then
separate or strain out and sparge hops. Chill the wort to 70
degrees F (21 C) and direct into a sanitized fermenter. Aerate the cooled wort well. Add an active yeast culture and
ferment for 4 to 6 days in the primary. Then transfer into a
secondary fermenter and age for one month, preferably at
about 70 degrees F (21 C).

When secondary aging is complete, prime with sugar, bottle or keg. Let condition at temperatures above 60 degrees F
(15.5 C) until clear and carbonated.

MASH-EXTRACT RECIPE AND PROCEDURE FOR PAUWEL KWAK
6   lbs.   (2.7 kg.)   English amber dried malt extract

1½ lbs. (0.7 kg.) French Pilsener malt or other Pilsener malt

3½ lbs. (1.6 kg.) Belgian Munich malt—7 Lovibond (15 EBC)

5   lbs. (2.3 kg.) Total grains

4   HBU (113 MBU) English Challenger hops (pellets)—60 minutes (bittering)

2.5 HBU (71 MBU) European Styrian Goldings hops (pellets)—15 minutes (flavor/aroma)

1   HBU Czech Saaz hops (pellets)—15 minutes (flavor/aroma)

¼   tsp. Irish moss

¾   c.   corn sugar for priming in bottles. Use 1/3 cup corn sugar if priming a keg.

Wyeast 1214 Belgian Ale yeast suggested; it is high-gravity- and ester-producing.

A step infusion mash is employed to mash the grains. Add 5 quarts (4.8 l.) of 140-degree F (60 C) water to the crushed grain, stir, stabilize and hold the temperature at 133 degrees F (56 C) for 30 minutes. Add 2.5 quarts (2.4 l.) of boiling water. Add heat to bring temperature up to 150 degrees F (65.5 C). Hold for about 60 minutes.

After conversion, raise temperature to 167 degrees F (75 C), lauter and sparge with 2 gallons (7.6 l.) of 170-degree F (77 C) water. Collect about 3 gallons (11.5 l.) of runoff. Add malt extract and bittering hops and bring to a full and vigorous boil.

The total boil time will be 60 minutes. When 15 minutes remain, add flavor hops and Irish moss. After a total wort boil of 60 minutes, turn off the heat, then separate or strain out and sparge hops, and direct the hot wort into a sanitized fermenter to which 2 gallons (7.6 l.) of cold water have been added. If necessary, add additional cold water to achieve a 5-gallon (19-l.) batch size. Chill the wort to 70 degrees F (21 C). Aerate the cooled wort well. Add an active yeast culture and ferment for 4 to 6 days in the primary. Then transfer into a secondary fermenter and age for one month, preferably at about 70 degrees F (21 C).

When secondary aging is complete, prime with sugar, bottle or keg. Let condition at temperatures above 60 degrees F (15.5 C) until clear and carbonated.

Silver Cup Winner

La Chouffe
Brasserie D'Achouffe
Achouffe, Belgium

Golden pale color with a slight chill haze. From the notably dense, rich head emerges all that the label indicates: 8 percent alcohol by volume, bottle conditioning, Pilsener malts, English Goldings and Saaz hops, coriander for spice. Full mellow fruitiness marries nicely with the floral nature of coriander spice. Fruitiness is a bit reminiscent of banana, but not overtly so. Goldings hop aroma is largely masked by the coriander, but evident if sought. Well-rounded medium- to full-bodied mouth feel. Fruity flavors with hints of banana and the zesty combination of coriander and alcohol. Because of so much malt fullness, the alcohol is not bitter or overt, simply warming in the aftertaste and the mind's eye. Bitterness is soft and gentle, balancing the bigness of this ale. Aftertaste is slightly bitter after the sweetness and floral character subside. All characters seem clear and absent of interference. A very clean beer.

Estimated profile based on tasting
Alcohol by volume: 8% as indicated on the label

Color: 5 SRM (10 EBC)
Bittering Units: 35–36

Bronze Cup Winner

None

CATEGORY 32:
BELGIAN-STYLE WHITE (OR WIT)/
BELGIAN-STYLE WHEAT

Belgian White Ales are brewed using unmalted and/or malted wheat and malted barley, and can be spiced with coriander and orange peel. These very pale beers are typically cloudy. The style is further characterized by the use of noble-type hops to achieve a low to medium bitterness and hop flavor. This dry beer has low to medium body, no diacetyl and a low fruity-ester content.

Original Gravity (°Plato):	1.044–1.050 (11–12.5 °Plato)
Apparent Extract– Final Gravity (°Plato):	1.006–1.010 (1.5–2.5 °Plato)
Alcohol by weight (volume):	3.8–4.4% (4.8–5.2%)
Bitterness (IBU):	15–25
Color SRM (EBC):	2–4 (5–10 EBC)

Gold Cup Winner

Hoegaarden White Beer
Brouwerij de Kluis
46 Stoopkens Straat

3320 Hoegaarden, Belgium
Brewmaster: Eddy Van Der Heggen
Established 1978 (give or take 400 years)
Production: 632,000 bbl. (740,000 hl.)

According to municipal archives, breweries existed in Hoegaarden by the year 1318. This was a land of people who always loved their beer. In the seventeenth century the southern provinces of the Netherlands (of which Hoegaarden is a part) boasted 3,223 breweries. In 1880 Hoegaarden supported 85 breweries with only 2,000 inhabitants! But times have changed. Now there are barely 100 breweries in all of Belgium, Brouwerij de Kluis being one of them.

In centuries past the most popular beer style of the region was known as wit beer, a white, cloudy, spicy brew, made from local barley malt and wheat. Its popularity played itself out, and along with the industrialization of the region, the

Town of Hoegaarden

last brewery closed its doors in Hoegaarden in 1957. It was the reminiscence of a local milkman that brought white beer back from its grave. Brewing from his kitchen with home-made equipment, the milkman, along with the encourage-ment of close and thirsty friends, set out making the fabled cloudy white beer he had learned to make by watching brew-masters when he was a boy. That was in 1965. By 1978 a soft-drink factory was converted into a large white beer brewery and the new Hoegaarden Brewery was on its leg-endary way. In 1985 a fire leveled the brewery, but it was soon rebuilt to continue reviving the appreciation for Bel-gian white beer throughout the world. Soon after the fire, the milkman sold the brewery to the large brewery In-terbrew, which has maintained the Hoegaarden white beer tradition since.

The milkman is Pierre Celis, who moved to Texas and established a Belgian-style American white beer brewery, The Celis Brewery.

Besides the award-winning Hoegaarden White Beer, the brewery also produces Hoegaarden Grand Cru (a stronger and more complex version of the White Beer), Forbidden Fruit (a strong, dark ale reminiscent of Grand Cru with roasted malt) and Julius (a blond all-malt strong ale).

CHARACTER DESCRIPTION OF GOLD CUP–WINNING HOEGAARDEN WHITE BEER

Extremely pale in color. Bottle-conditioned with yeast haze appropriately evident. Very exceptional fine, white, sat-iny head with great head retention. Citrusy sweet lemonlike floral aroma attributable to coriander and curaçao orange peel. Slight banana fruitiness from fermentation but cer-tainly not aggressive. Wonderful fresh, sassy and sexy yeast aroma. The floral notes of coriander, orange peel and herbal hops blend to become one complex alluring siren, clothed in sheer satin. First flavor impression finds the beer notably effervescent. Floral and fruity flavors blossom, but are not acidic. Full, sensual mouth feel without bitterness. Clean aftertaste refreshes.

- **Recipe for 5 U.S. gallons (19 liters) Hoegaarden White Beer**
  Targets:
  Original Gravity: 1.048 (12)
  Final Gravity: 1.010 (2.5)
  Alcohol by volume: 5%
  Color: 4 SRM (8 EBC)
  Bittering Units: 13

ALL-GRAIN RECIPE AND PROCEDURE

4½	lbs. (2 kg.)	Belgian 2-row Pilsener malt
4½	lbs. (2 kg.)	flaked unmalted wheat
9	**lbs. (3.2 kg.)**	**Total grains**

2.5 HBU (71 MBU) American Nugget hops (pellets)—75 minutes (bittering)

1 HBU (28 MBU) Czech Saaz hops (pellets)—15 minutes (flavor)

1.2 HBU (33 MBU) European Styrian Goldings hops (pellets)—15 minutes (flavor)

1 oz. (28 g.) freshly ground coriander seed

¼ oz. (7 g.) dried ground curaçao orange peel

¼ tsp. Irish moss

¾ c. corn sugar for priming in bottles. Use ⅓ cup corn sugar if priming a keg.

Wyeast 3944 Belgian Witbier yeast is recommended.

A step infusion mash is employed to mash the grains. Add 9 quarts (8.6 l.) of 130-degree F (54.5 C) water to the crushed grain, stir, stabilize and hold the temperature at 122 degrees F (50 C) for 30 minutes. Add 4.5 quarts (4.3 l.)

of boiling water. Add heat to bring temperature up to 150 degrees F (65.5 C). Hold for about 60 minutes.

After conversion, raise temperature to 167 degrees F (75 C), lauter and sparge with 4.5 gallons (17 l.) of 170-degree F (77 C) water. Collect about 6 gallons (23 l.) of runoff, add bittering hops and bring to a full and vigorous boil.

The total boil time will be 75 minutes. When 15 minutes remain, add flavor hops and Irish moss. When 5 minutes remain, add ½ ounce (14 g.) of coriander seed and ¼ ounce (7 g.) orange peel. After a total wort boil of 75 minutes (reducing the wort volume to just over 5 gallons), turn off the heat, then separate or strain out and sparge hops. Chill the wort to 70 degrees F (21 C) and direct into a sanitized fermenter. Aerate the cooled wort well. Add an active yeast culture and ferment for 4 to 6 days in the primary. Then transfer into a secondary fermenter and add remaining ½ ounce (14 g.) crushed coriander seed. Allow to age for two weeks.

When secondary aging is complete, prime with sugar, bottle or keg. Let condition at temperatures above 60 degrees F (15.5 C) until clear and carbonated.

MALT-EXTRACT RECIPE AND PROCEDURE FOR HOEGAARDEN WHITE BEER

4½ lbs.	(2 kg.)	wheat malt extract (50% wheat, 50% barley)
2¼ lbs.	(1 kg.)	very light honey
0 lb.	**(0 kg.)**	**Total grains**

3	HBU	(85 MBU) American Nugget hops (pellets)—60 minutes (bittering)
1	HBU	(28 MBU) Czech Saaz hops (pellets)—15 minutes (flavor)
1.2	HBU	(33 MBU) European Styrian Goldings hops (pellets)—15 minutes (flavor)
1	oz.	(28 g.) freshly ground coriander seed
¼	oz.	(7 g.) dried ground curaçao orange peel

¼   tsp.   Irish moss
¾   c.   corn sugar for priming in bottles. Use ⅓ cup
                corn sugar if priming a keg.
Wyeast 3944 Belgian Witbier yeast is recommended.

Add the malt extract and bittering hops to 2.5 gallons
(9.5 l.) of water and bring to a full and vigorous boil. The
total boil time will be 60 minutes. When 15 minutes remain,
add flavor hops and Irish moss. When 5 minutes remain,
add ½ ounce (14 g.) of coriander seed and ¼ ounce (7 g.)
orange peel. After a total wort boil of 60 minutes, turn off
the heat, separate or strain out and sparge hops, and direct
the hot wort into a sanitized fermenter to which 2 gallons
(7.6 l.) of cold water have been added. If necessary, add
additional cold water to achieve a 5-gallon (19 l.) batch size.
Chill the wort to 70 degrees F (21 C). Aerate the cooled
wort well. Add an active yeast culture and ferment for 4 to
6 days in the primary. Then transfer into a secondary fermenter and add remaining ½ ounce (14 g.) crushed coriander seed. Allow to age for two weeks.

When secondary aging is complete, prime with sugar, bottle or keg. Let condition at temperatures above 60 degrees
F (15.5 C) until clear and carbonated.

Silver Cup Winner

Wit
Spring Street Brewing Co.
New York, New York, USA

Light amber color. Intriguing and mystical toasted-malt
aroma accompanied by a touch of caramel lurking in the
background. The very refreshing and zesty aromatic character of coriander spice takes the spotlight and welcomes indulgence. Perhaps there is a soft noble hop aromatic, but
it may be the imagination after the strong performance of

other characters. Medium- to light-bodied mouth feel. A clean, dry finish and very little bitterness. The small amount of bitterness plays a supporting role to the other characters, contributing only to the balance. Taste is very smooth and clean. A soft, gentle yeast character contributes to the overall memory if sediment is poured with the beer. Perhaps there is a bit of orange-peel fruitiness, but in very reserved amounts. A unique beer, easy to drink, digest, enjoy and discuss.

Estimated profile based on tasting
Color: 5–6 SRM (10–12 EBC)
Bittering Units: 17–19

Bronze Cup Winner

Celis White
Celis Brewery Inc.
Austin, Texas, USA

Very pale, light straw color. Slightly hazy from yeast sediment and wheat. Notable foam clinging to glass. Aroma is actively complex and explosive. Fruity (very low in banana character), spicy and clovelike aroma with some coriander evident as well. Yeast sediment contributes to aroma and flavor. Citrus from orange peel and a mild clove flavor. Light body with a dry finish is associated with a complex and active flavor. A fun and enjoyable beer with very low bitterness. Fruitiness is slightly reminiscent of blueberries. Orange peel is accented as beer warms. Not to be served at very cold temperatures, otherwise you'll miss a lot of the complex character. An American classic in the Belgian tradition.

Estimated profile based on tasting
Color: 3–4 SRM (6–8 EBC)
Bittering Units: 20–22

CATEGORY 33:
BELGIAN-STYLE LAMBIC

A. SUBCATEGORY: BELGIAN-STYLE LAMBIC

These unblended, naturally fermented lambic beers are intensely estery, sour and acetic-flavored. They are low in CO_2. These haze beers are brewed with unmalted wheat and malted barley. They are quite dry and light-bodied and very low in hop bitterness. Cloudiness is acceptable.

Original Gravity (°Plato):	1.044–1.056 (11–14 °Plato)
Apparent Extract– Final Gravity (°Plato):	1.000–1.010 (0–2.5 °Plato)
Alcohol by weight (volume):	4–5% (5–6%)
Bitterness (IBU):	11–23
Color SRM (EBC):	6–13 (15–33 EBC)

B. SUBCATEGORY: BELGIAN-STYLE GUEUZE LAMBIC

These unflavored, blended and secondary-fermented lambic beers may be very dry or mildly sweet and are characterized by an intensely fruity-estery, sour and acidic flavor. These pale beers are brewed with unmalted wheat, malted barley and stale, aged hops. They are very low in hop bitterness. Cloudiness is acceptable. These beers are quite dry and light-bodied.

Original Gravity (°Plato):	1.044–1.056 (11–14 °Plato)
Apparent Extract– Final Gravity (°Plato):	1.000–1.010 (0–2.5 °Plato)
Alcohol by weight (volume):	4.0–5.0% (5.0–6.0%)

Bitterness (IBU):   11–23

Color SRM (EBC):   6–13 (15–33 EBC)

C. SUBCATEGORY: BELGIAN-STYLE FRUIT LAMBIC

These beers, also known by the names Framboise, Kriek, Peche and others, are characterized by fruit flavor and aroma. The intense color reflects the choice of fruit. Sourness dominates the flavor profile. These flavored lambic beers may be very dry or mildly sweet.

Original Gravity (°Plato):   1.040–1.072 (10–17.5 °Plato)

Apparent Extract–
Final Gravity (°Plato):   1.008–1.016 (2–4 °Plato)

Alcohol by weight (volume): 4.0–5.5% (5.0–7.0%)

Bitterness (IBU):   15–21

Color SRM (EBC):   NA. Light color takes on hue of fruit.

Gold Cup Winner

Lindemans Cuvée René Grand-Cru Gueuze Lambic
Lindemans Brewery
257 Lenniksebaan
1712 Vlezenbeek, Belgium
Brewmaster: René Lindeman
Established 1811
Production: 21,000 bbl. (25,000 hl.)

Lindemans Brewery is a very real farmhouse brewery on the outskirts of Brussels, in the town of Vlezenbeek. It began producing beer for sale in 1811, but long before that it produced beer for farm and friends. The brewery is a marvel

Lindemans Brewery in Vlezenbeek, Belgium

of tradition and uniqueness, as are all lambic breweries. Lindemans has the distinction of being the first brewery to import Gueuze (pronounced GUZ-ah) and fruit lambics to the United States, through Merchant du Vin of Seattle, Washington, in the mid-1970s.

The spontaneously fermented Gueuze style of beer is indigenous to the area in and around Brussels, for only in this region of the world do the requisite microorganisms thrive in the air. Traditionally brewed during the cooler months of the year when the balance of airborne microorganisms is perfect, lambic beers are begun as a wort based on about 50 percent malted barley and 50 percent unmalted wheat. Hops used are typically aged for over three years at room temperature. After the wort is produced and cooled in a large, open, shallow vessel, the rafters are opened and airborne yeasts and bacteria are allowed to settle and thrive in the wort. Spontaneous fermentation begins and the beer is cared for in fermentation tanks and traditional oak barrels for several years before considered matured.

Cuvée René is a blend of one part older lambics (one and a half to two years old) with two parts young lambic (six months to one year old). The unfermented sugars in the young beer induce a refermentation in this truly remarkable bottle-conditioned beer. Still imported by pioneer beer importer Merchant du Vin, other beers produced at Lindemans include Framboise (raspberry) Lambic, Peche (peach) Lambic, Gueuze Lambic, Kriek (cherry) Lambic and Faro Lambic.

CHARACTER DESCRIPTION OF GOLD CUP–WINNING LINDEMANS
CUVÉE RENÉ CRAND-CRU GUEUZE LAMBIC

Bottle-conditioned and joyously effervescent. Golden color with a light orange hue. Rich head with very good head retention. Aroma explodes with a big, powerful, fruity, estery, acidic, earthy-lambic character, possessing all the traits of a well-blended Gueuze. A classic. Flavor is a very grapefruitlike sour acidity with a reassuring earthy, musty foundation. The mouth feel is extraordinarily dry in the true tradition of lambics and is not sweetened with sugar just before bottling (and then pasteurized after bottling), as is typical of some other brands of Gueuze. The aftertaste leaves the palate refreshed and cleansed after the initial assault of acidity. Best enjoyed reasonably chilled as higher solventlike alcohols emerge as the beer warms.

No one can take this piece of the brewer's art away from the brewery. It is like a poem, inspired and original and not duplicable. But we can try.

- **Recipe for 5 U.S. gallons (19 liters) Lindemans Cuvée René Crand-Cru Gueuze Lambic**
  Targets:
  Original Gravity: 1.050 (12.5)
  Final Gravity: 1.000 (0.05)
  Alcohol by volume: 6%
  Color: 6 SRM (12 EBC)
  Bittering Units: 6

ALL-GRAIN RECIPE AND PROCEDURE

3¼   lbs. (1.5 kg.)   flaked unmalted wheat
5¾   lbs. (2.6 kg.)   Belgian 2-row pale ale malt
9   lbs. (4.1 kg.)   Total grains

Traditionally hops used are aged at room temperature for three to four years. Lindemans ages American Northern Brewer and uses them at an approximate rate of 4 ounces (113 g.) of whole hops per 5 gallons (19 l.). Equivalent HBU would be about 5 units of HBUs to achieve the desired low level of bitterness.

4   oz.   (113 g.) 3-year-old aged American Northern
              Brewer hops (whole)—60 minutes (bittering)
or
5   HBU (137 MBU) American Northern Brewer hops
             (whole)—75 minutes (bittering)

¼   tsp.   Irish moss
¾   c.   corn sugar for priming in bottles. Use ⅓ cup
             corn sugar if priming a keg.
Wyeast 3278 Belgian Lambic blend yeast

A step infusion mash is employed to mash the grains. Add 9 quarts (8.6 l.) of 130-degree F (54.5 C) water to the crushed grain, stir, stabilize and hold the temperature at 122 degrees F (50 C) for 30 minutes. Add 4.5 quarts (4.3 l.) of boiling water. Add heat to bring temperature up to 150 degrees F (65.5 C). Hold for about 60 minutes.

After conversion, raise temperature to 167 degrees F (75 C), lauter and sparge with 4.5 gallons (17 l.) of 170-degree F (77 C) water. Collect about 6 gallons (23 l.) of runoff, add bittering hops and bring to a full and vigorous boil.

The total boil time will be 75 minutes. When 15 minutes remain, add Irish moss. After a total wort boil of 75 minutes (reducing the wort volume to just over 5 gallons), turn off the heat, then separate or strain out and sparge hops. Chill the wort to 70 degrees F (21 C) and direct into a sanitized fermenter. Aerate the cooled wort well. Add an active yeast culture and ferment for one month in the primary. Then transfer to a secondary and let age undisturbed for one to two years. An ugly but reassuring skin of mold should develop on the surface. Do not disturb this protective film. Age at room temperatures, 60 to 70 degrees F (15.5–21 C).

Brew a batch every year, and as the beers age, remove a small portion for tasting. Blend as you choose or bottle a single batch without blending. For added character, age in a used sherry oak barrel. Top off the barrel with aging lambic from other batches (perhaps aged in glass carboys) to compensate for evaporative losses through the porous oak.

When bottling a blend or a single one- to two-year-old batch, add priming sugar and a fresh dose of active ale yeast to assure carbonation.

Your rendition of Cuvée René will be unique. Be gloriously surprised and take much pride if you can come close to matching the character of Cuvée René.

MASH-EXTRACT RECIPE AND PROCEDURE FOR LINDEMANS CUVÉE RENÉ GRAND-CRU GUEUZE LAMBIC

2	lbs.	(0.9 kg.)	English extralight dried malt extract
2½	lbs.	(1.1 kg.)	flaked unmalted wheat
3½	lbs.	(1.6 kg.)	Belgian 2-row pale ale malt
6	**lbs.**	**(2.7 kg.)**	**Total grains**

Traditionally hops used are aged at room temperature for three to four years. Lindemans ages American Northern Brewer and uses them at an approximate rate of 4 ounces (113 g.) of whole hops per 5 gallons (19 l.). Equivalent HBU would be about 5 units of HBUs to achieve the desired low level of bitterness.

| 4 | oz. | (113 g.) 3-year-old aged American Northern Brewer hops (whole)—60 minues (bittering) |

or

5   HBU (137 MBU) American Northern Brewer hops (whole)—75 minutes (bittering)

¼ tsp.   Irish moss
¾ c.   corn sugar for priming in bottles. Use ⅓ cup corn sugar if priming a keg.
Wyeast 3278 Belgian Lambic blend yeast

A step infusion mash is employed to mash the grains. Add 6 quarts (5.7 l.) of 138-degree F (58 C) water to the crushed grain, stir, stabilize and hold the temperature at 128 degrees F (53 C) for 30 minutes. Add 3 quarts (2.9 l.) of boiling water. Add heat to bring temperature up to 150 degrees F (65.5 C). Hold for about 60 minutes.

After conversion, raise temperature to 167 degrees F (75 C), lauter and sparge with 2 gallons (7.6 l.) of 170-degree F (77 C) water. Collect about 3 gallons (11.5 l.) of runoff. Add malt extract and bittering hops and bring to a full and vigorous boil.

The total boil time will be 60 minutes. When 15 minutes remain, add Irish moss. After a total wort boil of 60 minutes (reducing the wort volume to just over 5 gallons), turn off the heat, then separate or strain out and sparge hops, and direct the hot wort into a sanitized fermenter to which 2 gallons (7.6 l.) of cold water have been added. If necessary, add additional cold water to achieve a 5-gallon (19-l.) batch size. Chill the wort to 70 degrees F (21 C). Aerate the cooled wort well. Add an active yeast culture and ferment for one month in the primary. Then transfer to a secondary and let age undisturbed for one to two years. A skin of mold should develop on the surface. Do not disturb this protective film. Age at room temperatures, 60 to 70 degrees F (15.5–21 C).

Brew a batch every year, and as the beers age, remove a small portion for tasting. Blend as you choose or bottle a single batch without blending. For added character, age in a used sherry oak barrel. Top off the barrel with aging lambic from other batches (perhaps aged in glass carboys) to compensate for evaporative losses through the porous oak.

When bottling a blend or a single one- to two-year-old batch, add priming sugar and a fresh dose of active ale yeast to assure carbonation.

Your rendition of Cuvée René will be unique. Be gloriously surprised and take much pride if you can come close to matching the character of Cuvée René.

Silver Cup Winner

None

Bronze Cup Winner

Belle-Vue Kriek
Brasserie Belle-Vue
Brussels, Belgium

Clear, deep, ruby red; exotic in its transparency. Robust and earthy cherry aroma reminiscent of wild cherry but not quite as sweet. Exotic earthy and musty aromas (like a damp, dirt-floored cellar) equally as memorable as the anticipated flavor to follow. Full-bodied mouth feel complements the intense cherry flavor. Just enough acidity balances the cherry flavor between sourness and sweetness. No malt character or hop flavors noted. Earthy, musty, spicy, almost cedarlike flavors abound throughout the experience as a backdrop to the high notes of fruitiness. Despite the original fullness in the mouth feel, there is a clean, dry finish due to the balance of acidity and tannin from the cherries. Astringency from cherries is minimal, the beer having been aged three years in oak barrels. A small hint of bitterness comes through in the aftertaste.

Estimated profile based on tasting
Color: Red; 22–27 SRM (44–54 EBC)
Bittering Units: 10–12

CATEGORY 34:
FRENCH-STYLE BIÈRE DE GARDE

Beers in this category are golden to deep copper or light brown. They are light to medium in body. The beer is characterized by a malty, often toasted-malt aroma, slight malt sweetness and medium hop bitterness. Noble-type hop aromas and flavors should be low to medium. Fruity esters can be light to medium in intensity. Earthy, cellarlike, musty aromas are acceptable. Diacetyl should not be perceived, but chill haze is okay. This style is often bottle-conditioned with some yeast character.

Original Gravity (°Plato):	1.060–1.080 (15–20 °Plato)
Apparent Extract– Final Gravity (°Plato):	1.012–1.024 (3–6 °Plato)
Alcohol by weight (volume):	3.5–6.3% (4.5–8%)
Bitterness (IBU):	25–30
Color SRM (EBC):	8–12 (16–30 EBC)

Gold Cup Winner

Grain d'Orge
Brasserie Jeanne d'Arc
38, Rue Anatole Rance
59790 Ronchin, France
Brewmaster: René LeBec
Established 1898
Production: 85,500 bbl. (100,000 hl.)

Known as the Flandres region, northern France has a rich brewing tradition dating back centuries. Charlemagne granted a brewing monopoly to French monks, and Louis IX proclaimed rules for the brewing trade in the thirteenth century. Intricately woven into the northern way of life, beer flows daily at the local bar, or *estaminet*. Adding to the sensuality, dozens of styles of beer accompany foods of the region that are created

René LeBec, Brewmaster

Simon Berdugo

with beer as an ingredient. Pils beer, special strong beers (usually over 6 percent alcohol), bières de garde and white beers, along with specialty seasonal beers, are the more popular beers of the area.

Inspired by the quality of the water in the region, Mr. Vandamme created the farmhouse Brasserie Jeanne d'Arc in 1898 in the small village of Ronchin, located near the French Flandres capital city of Lille. Homebrewing beer enthusiasts should note that the term *farmhouse brewery* has the word *house* in it. Brasserie Jeanne d'Arc was established on the premise of original homebrewed beer, as were many breweries of northern France.

Good business and marketing help support a brewery, but it's the beer that people ultimately support. The quality of Brasserie Jeanne d'Arc's beers allowed the brewery to expand its operations to include several bars and restaurants in the area, all serving its original creations. Though it currently produces lager beer brands such as Cristalor Alsatia, Pilsor, Orpal and Gold Triumph, the emphasis is on a tradition of producing indigenous and special beers, both for its regional and export markets. Embracing modernization

while maintaining the tradition is a skill that many breweries will need to learn if they are to survive. Apparently
Jeanne d'Arc has achieved a balance, offering such special
beers as Scotch Triumph, Bière de Noel, Bière de Mars,
Ambre des Flandres and this award-winning Grain d'Orge.

Brasserie Jeanne d'Arc is one of eighteen midsize breweries in France, all of which produce less than ten percent of
French beer.

CHARACTER DESCRIPTION OF GOLD CUP–WINNING GRAIN D'ORGE

As do many French bières de garde, Grain d'Orge intro-
duces itself in a corked bottle, virtually assuring a pleasant
earthy, musty aroma. Appearance is crystal-clear and light
amber in color. Well carbonated with a pleasant head. First
aromatic impressions are a gentle, mild maltiness and
musty cork character. Subdued ale aromatics contribute to
an easy complexity without fruity esters. Full-bodied with a
malty finish. The assault of malt and alcohol character on
the palate highlights the beer's overall impact. Noble hop
bitterness is soft and mild and generally balanced toward
neutrality; hop flavor does not emerge. Pale malts with a
hint of Munich and CaraMunich-type malts seem apparent,
but the formulation is actually aromatic Vienna and French
caramel malt.

- **Recipe for 5 U.S. gallons (19 liters) Grain d'Orge**
  Targets:
  Original Gravity: 1.087 (210)
  Final Gravity: 1.026 (6.5)
  Alcohol by volume: 8%
  Color: 8 SRM (16 EBC)
  Bittering Units: 20

ALL-GRAIN RECIPE AND PROCEDURE

9   lbs. (4.1 kg.)   French 6-row Gatinais F or Belgium
                       Pilsener malt

½	lb.	(0.23 kg.)	French Aromatic Vienne—3.5 Lovibond (7–8 EBC) or Vienna malt
¼	lb.	(114 g.)	French caramel malt—20 Lovibond (50 EBC) or crystal malt—20 Lovibond
¼	lb.	(114 g.)	French caramel malt—30 Lovibond (60 EBC) or crystal malt—30 Lovibond
4¾	lbs.	(2.2 kg.)	flaked corn
1	lb.	(0.45 kg.)	sucrose
14¾ lbs.		**(6.7 kg.)**	**Total grains**

2	HBU	(56 MBU) German Hallertauer hops (pellets)—105 minutes (bittering)
2	HBU	(56 MBU) French or European Brewers Gold hops (pellets)—105 minutes (bittering)
2	HBU	(56 MBU) French or European Brewers Gold hops (pellets)—30 minutes (flavor)
1.5	HBU	(43 MBU) Slovenian Styrian Goldings hops (pellets)—10 minutes (aroma)

¼	tsp.	Irish moss
¾	c.	corn sugar for priming in bottles. Use ⅓ cup corn sugar if priming a keg.

Wyeast 1728 Scottish Ale yeast or other yeast producing a malty profile with low ester production and suitable for high-gravity fermentation.

A step infusion mash is employed to mash the grains. Add 15 quarts (14 l.) of 130-degree F (54.5 C) water to the crushed grain, stir, stabilize and hold the temperature at 122 degrees F (50 C) for 30 minutes. Add 7.5 quarts (7 l.) of boiling water, add heat to bring temperature up to 155 degrees F (68 C) and hold for about 60 minutes.

After conversion, raise temperature to 167 degrees F (75 C), lauter and sparge with 4 gallons (15 l.) of 170-degree F (77 C) water. Collect about 7 gallons (23 l.) of runoff, add sucrose and bittering hops, and bring to a full and vigorous boil.

The total boil time will be 105 minutes. When 30 minutes remain, add flavor hops. When 10 minutes remain, add aroma hops and Irish moss. After a total wort boil of 105 minutes (reducing the wort volume to just over 5 gallons), turn off the heat, then separate or strain out and sparge hops. Chill the wort to 70 degrees F (21 C) and direct into a sanitized fermenter. Aerate the cooled wort well. Add an active yeast culture and ferment for 4 to 6 days in the primary. Then transfer into a secondary fermenter, chill to 60 degrees F (15.5 C) if possible and allow to age for four weeks or more.

When secondary aging is complete, prime with sugar, bottle or keg. Let condition at temperatures above 60 degrees F (15.5 C) until clear and carbonated. Use corks and wire down the closure for musty-earthy "cork" character.

MASH-EXTRACT RECIPE AND PROCEDURE FOR GRAIN D'ORGE

4¼	lbs.	(1.9 kg.)	English extralight dried malt extract
4	lbs.	(1.8 kg.)	French 6-row Gatinais F or Belgium Pilsener malt
½	lb.	(0.23 kg.)	French aromatic Vienne—3.5 Lovibond (7–8 EBC) or Vienna malt
¼	lb.	(114 g.)	French caramel malt—20 Lovibond (50 EBC) or crystal malt—20 Lovibond
¼	lb.	(114 g.)	French caramel malt—30 Lovibond (60 EBC) or crystal malt—30 Lovibond
2½	lbs.	(1.1 kg.)	flaked corn
1	lb.	(0.45 kg.)	sucrose
7½	**lbs.**	**(3.4 kg.)**	**Total grains**

2.5 HBU (71 MBU) German Hallertauer hops (pellets)— 90 minutes (bittering)

3 HBU (85 MBU) French or European Brewers Gold hops (pellets)—90 minutes (bittering)

2 HBU (56 MBU) French or European Brewers Gold hops (pellets)—30 minutes (flavor)

1.5 HBU (43 MBU) Slovenian Styrian Goldings hops (pellets)—10 minutes (aroma)

¼ tsp. Irish moss

¾ c. corn sugar for priming in bottles. Use ⅓ cup corn sugar if priming a keg.

Wyeast 1728 Scottish Ale yeast or other yeast producing a malty profile with low ester production and suitable for high-gravity fermentation.

A step infusion mash is employed to mash the grains. Add 7.5 quarts (7 l.) of 130-degree F (54.5 C) water to the crushed grain, stir, stabilize and hold the temperature at 122 degrees F (50 C) for 30 minutes. Add 4 quarts (2.9 l.) of boiling water, add heat to bring temperature up to 155 degrees F (68 C) and hold for about 60 minutes.

After conversion, raise tempertaure to 167 degrees F (75 C), lauter and sparge with 2.5 gallons (9.5 l.) of 170-degree F (77 C) water. Collect about 4.5 gallons (17 l.) of runoff. Add malt extract, sucrose and bittering hops, and bring to a full and vigorous boil.

The total boil time will be 90 minutes. When 30 minutes remain, add flavor hops. When 10 minutes remain, add aroma hops and Irish moss. After a total wort boil of 90 minutes, turn off the heat, then separate or strain out and sparge hops, and direct the hot wort into a sanitized fermenter to which 2 gallons (7.6 l.) of very cold water have been added. If necessary, add additional cold water to achieve a 5-gallon (19-l.) batch size. Chill the wort to 70 degrees F (21 C). Aerate the cooled wort well. Add an active yeast culture and ferment for 4 to 6 days in the primary. Then transfer into a secondary fermenter, chill to 60 degrees

F (15.5 C) if possible and allow to age for four weeks or more.

When secondary aging is complete, prime with sugar, bottle or keg. Let condition at temperatures above 60 degrees F (15.5 C) until clear and carbonated. Use corks and wire down the closure for musty-earthy "cork" character.

Silver Cup Winner

None

Bronze Cup Winner

None

Lager Beers

European-Germanic Origin

CATEGORY 35:
GERMAN-STYLE PILSENER

A classic German Pilsener is very light straw/golden-colored and well hopped. Hop bitterness is high. Hop aroma and flavor are moderate and quite obvious. It is a well-attenuated, medium-bodied beer, but a malty accent can be perceived. Fruity esters and diacetyl should not be perceived. There should be no chill haze. Its head should be dense and rich.

Original Gravity (°Plato):   1.044–1.050 (11–12.5 °Plato)
Apparent Extract–
Final Gravity (°Plato):   1.006–1.012 (1.5–3 °Plato)
Alcohol by weight (volume): 3.6–4.2% (4–5%)

Bitterness (IBU):   30–40
Color (SRM):   3–4 (7–10 EBC)

Gold Cup Winner

Redwood Coast Alpine Gold Pilsner
Redwood Coast Brewing Company
Pacific Marina 1051
Alameda, California, USA 94501
Brewmaster: Dr. Ronald Manabe
Established 1987
Production: 4,500 bbl. (5,300 hl.)

Redwood Coast also won the English-style Brown Ale category. Please see that category for a description of the brewery.

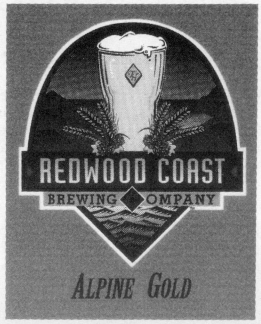

Ronald M. Manabe

Very pale straw gold color with rich head formation. Tantalizing aroma is gently powered with the herbal-minty character of European noble hops of the Hallertauer genre. Appropriately ester-free. Cold and well aged. Malt character is at very low levels, making the crisp character of a classic Pilsener apparent. Flavor is a perfect adaptation of the classic quality German Pils. Good hop bite, but with a soft bitterness. Dry and refreshing first impression. Light to barely medium body, finishing dry and with a clean bitterness in the aftertaste, which is both quenching and seductive. Perfectly balanced between bitterness and sweetness; very little malt sweetness comes through. Aftertaste is clean and softly bitter with a warmth of alcohol felt in mouth/aftertaste, asking for further indulgence. Yes, please. Thank you.

- **Recipe for 5 U.S. gallons (19 liters) Redwood Coast Alpine Gold Pilsner**
Targets:
Original Gravity: 1.049 (12.5)
Final Gravity: 1.010 (2.5)
Alcohol by volume: 5%
Color: 5 SRM (10 EBC)
Bittering Units: 30

ALL-GRAIN RECIPE AND PROCEDURE

8 lbs. (3.6 kg.) American/English 2-row Klages/Harrington pale malt
¼ lb. (114 g.) American wheat malt
½ lb. (0.23 kg.) American Munich malt
¼ lb. (114 g.) American Cara-Pils malt
9 lbs. (3.2 kg.) Total grains

3.5 HBU (99 MBU) German Hersbrucker Hallertauer hops (pellets)—75 minutes (bittering)

0.5	HBU	(14 MBU) Czech Saaz hops (pellets)—75 minutes (bittering)
1.2	HBU	(33 MBU) German Hersbrucker Hallertauer hops (pellets)—30 minutes (flavor)
1	HBU	(28 MBU) Czech Saaz hops (pellets)—30 minutes (flavor)
2.5	HBU	(71 MBU) German Hersbrucker Hallertauer hops (pellets)—10 minutes (aroma)
1	HBU	(28 MBU) Czech Saaz hops (pellets)—10 minutes (aroma)

| ¼ | tsp. | Irish moss |
| ¾ | c. | corn sugar for priming in bottles. Use ⅓ cup corn sugar if priming a keg. |

Wyeast 2278 Czech Pils yeast or Weihenstephan 34–70. A lager yeast producing a crisp finish, low malt character, good attenuation and no esters, sulfur or diacetyl is recommended.

A step infusion mash is employed to mash the grains. Add 9 quarts (8.5 l.) of 140-degree F (60 C) water to the crushed grain, stir, stabilize and hold the temperature at 133 degrees F (56 C) for 30 minutes. Add 4.5 quarts (4.3 l.) of boiling water, adding heat if necessary to bring temperature up to 150 degrees F (65.5 C). Hold for about 60 minutes.

After conversion, raise temperature to 167 degrees F (75 C), lauter and sparge with 4 gallons (15 l.) of 170-degree F (77 C) water. Collect about 6 gallons (23 l.) of runoff, add bittering hops and bring to a full and vigorous boil.

The total boil time will be 75 minutes. When 30 minutes remain, add flavor hops. When 10 minutes remain, add aroma hops and Irish moss. After a total wort boil of 75 minutes (reducing the wort volume to just over 5 gallons), turn off the heat, then separate or strain out and sparge hops. Chill the wort to 65 degrees F (18 C) and direct into a sanitized fermenter. Aerate the cooled wort well. Add an active yeast culture and ferment for 4 to 6 days in the primary at 55 degrees F (15 C). Then transfer into a secondary fermenter, chill to 50 degrees F (10 C) to age for two more weeks, then lager for two to four more weeks at 40 degrees F (4.5 C).

When secondary aging is complete, prime with sugar, bottle or keg. Let condition at temperatures above 60 degrees F (15.5 C) until clear and carbonated, then store chilled.

MASH-EXTRACT RECIPE AND PROCEDURE FOR REDWOOD COAST ALPINE GOLD PILSNER

4	lbs. (1.8 kg.)	English extralight dried malt extract
2	lbs. (0.9 kg.)	American/English 2-row Klages/Harrington pale malt
¼	lb. (114 g.)	American wheat malt
½	lb. (0.23 kg.)	American Munich malt
¼	lb. (114 g.)	American Cara-Pils malt
3	**lbs. (1.4 kg.)**	**Total grains**

4	HBU	(113 MBU) German Hersbrucker Hallertauer hops (pellets)—60 minutes (bittering)
1	HBU	(28 MBU) Czech Saaz hops (pellets)—60 minutes (bittering)
4	HBU	(113 MBU) German Hersbrucker Hallertauer hops (pellets)—30 minutes (flavor)
1	HBU	(28 MBU) Czech Saaz hops (pellets)—30 minutes (flavor)
2.5	HBU	(71 MBU) German Hersbrucker Hallertauer hops (pellets)—10 minutes (aroma)
1	HBU	(28 MBU) Czech Saaz hops (pellets)—10 minutes (aroma)

¼	tsp.	Irish moss
¾	c.	corn sugar for priming in bottles. Use ⅓ cup corn sugar if priming a keg.

Wyeast 2278 Czech Pils yeast or Wyeast 34–70. A lager yeast producing a crisp finish, low malt character, good attenuation and no esters, sulfur or diacetyl is recommended.

A step infusion mash is employed to mash the grains. Add 3 quarts (3 l.) of 140-degree F (60 C) water to the crushed grain, stir, stabilize and hold the temperature at 133 degrees F (56 C) for 30 minutes. Add 1.5 quarts (1.5 l.) of boiling water, adding heat if necessary to bring temperature up to 150 degrees F (65.5 C). Hold for about 60 minutes.

After conversion, raise temperature to 167 degrees F (75 C), lauter and sparge with 2 gallons (7.6 l.) of 170-degree F (77 C) water. Collect about 3 gallons (11.5 l.) of runoff. Add malt extract and bittering hops and bring to a full and vigorous boil.

The total boil time will be 60 minutes. When 30 minutes remain, add flavor hops. When 10 minutes remain, add aroma hops and Irish moss. After a total wort boil of 60 minutes, turn off the heat, then separate or strain out and sparge hops, and direct the hot wort into a sanitized fermenter to which 2 gallons (7.6 l.) of cold water have been added. If necessary, add additional cold water to achieve a 5-gallon (19-l.) batch size. Chill the wort 65 degrees F (68.5 C). Aerate the cooled wort well. Add an active yeast culture and ferment for 4 to 6 days in the primary at 55 degrees F (15 C). Then transfer into a secondary fermenter, chill to 50 degrees F (10 C) to age for two more weeks, then lager for two to four more weeks at 40 degrees F (4.5 C).

When secondary aging is complete, prime with sugar, bottle or keg. Let condition at temperatures above 60 degrees F (15.5 C) until clear and carbonated, then store chilled.

Silver Cup Winner

Bohemia Beer
Cerveceria Cuauhtèmoc
Monterrey, Mexico

Extremely pale straw color. Intriguing spicy Saaz hop aroma. Sweet aroma of malt shines along with the hop sparkle.

Aroma tends to drift towards fruitiness, but not from fermentation esters. Mouth feel is light-bodied. Hop flavor is an integral component, while gentle bitterness is not assertive. Bitterness and hop flavor complexly woven with malt during indulgence and in aftertaste. Clean, light aftertaste with nice lingering, clean, soft bitterness.

Estimated profile based on tasting
Color: 3–4 SRM (6–8 EBC)
Bittering Units: 25–28

Bronze Cup Winner

McClintic Pilsner
B.T. McClintic Brewing Co.
Janesville, Wisconsin, USA

Golden amber in color. Underlying malt aroma complemented with a hint of noble hops. Light-bodied beer with a dry finish. A neutral balance exists between malt and hops; sweet and bitter. No distinctive hop flavor, bitterness or aroma. Bitterness is there, but not a memorable character. Overall balance is on the bitter end, but the beer is dry and light, not requiring much hop bitterness to lend this impression. Light and refreshing, though generally lacking the body and overall texture for the style. Extraordinarily clean. Light-bodied. Could be more appropriately labeled as an American premium lager.

Estimated profile based on tasting
Color: 5–6 SRM (10–12 EBC)
Bittering Units: 18–22

CATEGORY 36:
BOHEMIAN-STYLE PILSENER

Pilseners in this subcategory are similar to German Pilseners; however, they are slightly more full-bodied and can be as dark as light amber. This style balances moderate bitterness and noble hop aroma and flavor with a malty, slightly sweet medium body. Diacetyl may be perceived in very low amounts. There should be no chill haze. Its head should be dense and rich.

Original Gravity (°Plato):	1.044–1.056 (11–14 °Plato)
Apparent Extract–	
Final Gravity (°Plato):	1.014–1.020 (3.5–5 °Plato)
Alcohol by weight (volume):	3.2–4% (4–5%)
Bitterness (IBU):	35–45
Color (SRM):	3–5 (7–14 EBC)

Gold Cup Winner

Ruffian Pilsner
Mountain Valley Brew Pub
122 Orange Ave.
Suffern, New York, USA 10901
Brewmaster: Jay Misson
Established 1992
Production: 4,000 bbl. (4,700 hl.)

Primarily designed for producing English-style ales using European malt and hops, Mountain Valley opened in 1992. Their Ruffian Porter took the Gold Medal in the Porter category at the 1994 Great American Beer Festival. The brewery's success continued and growth followed. In November of 1995 a 20-barrel decoction brewhouse was installed. Now

Ruffian Lagers are brewed following true German traditions, in terms of ingredients, equipment and process. Due to space limitations, brewmaster Jay Misson can only brew lagers occasionally. His first two attempts with their new brewhouse and yeast produced World Cup Gold winners. No small feat for Mountain Valley's home-brewer-turned-professional-brewmaster.

Jay enjoys relating the story of their first Hefe Weiss. The scheduled brew day was approaching, but their special yeast had not

Lon M. Lauterio

Jay Misson, Brewmaster

arrived. Fate and the brewery's guardian spirit became involved. A former brewery apprentice arrived for a visit the day before the brew with a bottle of Hefe Weiss from the brewery in Germany where Jay had worked. The yeast was cultured overnight and most of the following day and subsequently pitched into 15 barrels of wort. "Right on target! A true homebrewer technique that proved . . . fruitful," claims Jay.

What are some common homebrew techniques worth paying attention to when brewing this beer? Jay advises developing your kraeusening, mash acidification, dry-hopping and decoction-mashing techniques to approach the quality of his brews.

The 100-seat restaurant and tavern in Suffern, New York (thirty to forty-five minutes north of New York City), is open year-round. Any day is worth celebrating at the Mountain Valley Brew Pub, but special events such as Mai-Bach Fest in May, Summerfest in July, Oktoberfest in September and a St. Patrick's Day celebration in March create special experiences for those who visit.

With the vision of brewing traditional beers with only malt, hops, yeast and water, Pale Ale, Copper Ale, Hefe Weiss, and award-winning Porter, Pilsener and Mai Bock are among the several selections you'll find on tap and in bottles.

CHARACTER DESCRIPTION OF GOLD CUP–WINNING RUFFIAN PILSNER

Straw gold in color with suggestions of an amber-orange hue. Gentle herbal aroma of German-grown noble hops. Malt character not aggressive but pleasantly evident, serving as a foundation for the flavor overtones that follow. Hop bitterness is a skillful blend of sensations with a punctual, sharp, full-mouth bitterness that does not have a harsh impact, mellowing to a soft overall hop character. Hop character continues in the aftertaste, lingering delicately in the mouth and throat. Well balanced with medium creamy body that balances the hop character. The minimal sweetness of malt is expressed more in mouth feel than in flavor. Delectably drinkable. I'll have another.

- ### *Recipe for 5 U.S. gallons (19 liters) Ruffian Pilsner*
  Targets:
  Original Gravity: 1.048 (12)
  Final Gravity: 1.014 (3.5)
  Alcohol by volume: 5%
  Color: 7 SRM (14 EBC)
  Bittering Units: 35

ALL-GRAIN RECIPE AND PROCEDURE

8¼	lbs.	(3.7 kg.)	German Pilsener malt
½	lb.	(0.23 kg.)	German Munich malt
3.2	oz.	(90 g.)	German Sauer (sour) malt
9	**lbs.**	**(4.1 kg.)**	**Total grains**

8.8	HBU	(241 MBU) German Hersbrucker Spalt hops (whole)—90 minutes (bittering)
½	oz.	(14 g.) Czech Saaz hops (whole)—steep in finished boiled wort for 2 to 3 minutes (aroma)
¼	tsp.	Irish moss
¾	c.	corn sugar for priming in bottles. Use ⅓ cup corn sugar if priming a keg.

Wyeast 2278 Czech Pils yeast or other lager yeast producing a crisp finish, low malt character and good attenuation with no esters, sulfur or diacetyl.

A double-decoction mash is employed to mash the grains. If you are using alkaline or hard water, treat the water with an appropriate amount of lactic acid or refer to pages 120 to 122 of *The Home Brewer's Companion* for proceeding with a triple decoction with an acid rest.

Add 9 quarts (8.6 l.) of boiling water to the grain, raising the temperature to 122 degrees F (50 C). Then remove 4.5 quarts (4.3 l.) of the thickest mash and boil this in another vessel for 30 minutes. Stir this boiled decoction constantly. Maintain the remainder of the mash at 122 degrees F (50 C) while boiling the decocted portion. After boiling for 30 minutes, add the decoction back into the main mash vessel, raising the temperature to 150 degrees F (65.5). Maintain this temperature for one hour. Up to 9 quarts (8.6 l.) of boiling water may be added at any time during this period to maintain mashing temperature. After one hour remove one third of the mash (a fifty-fifty blend of thick mash and liquid) and boil about 20 minutes. Stir this second decoction constantly during the boil. When finished, return the boiled decoction to the main mash vessel, ending starch conversion by having raised the temperature to about 167 degrees F (75 C). Then lauter and sparge with 4 gallons (15 l.) of 170-degree F (77 C) water. Collect about 6 gallons (23 l.) of runoff, add bittering hops and bring to a full and vigorous boil.

The total boil time will be 90 minutes. When 10 minutes remain, add Irish moss. After a total wort boil of 90 minutes (reducing the wort volume to just over 5 gallons), turn off the heat, then add aroma hops and let steep 2 to 5 minutes. Then separate or strain out and sparge hops. Chill the wort to 60 degrees F (15.5 C) and direct into a sanitized fermenter. Aerate the cooled wort well. Add an active yeast culture and ferment for 4 to 6 days in the primary at about 55 degrees F (13 C). Then transfer into a secondary fermenter, chill to 50 degrees F (10 C) to age for two more weeks, then lager for two to four more weeks at 40 degrees F (4.5 C).

When secondary aging is complete, prime with sugar, bottle or keg. Let condition at temperatures above 60 degrees F (15.5 C) until clear and carbonated, then store chilled.

MASH-EXTRACT RECIPE AND PROCEDURE FOR RUFFIAN PILSNER

3¾	lbs. (1.7 kg.)	English extralight dried malt extract
2¼	lbs. (1 kg.)	German Pilsener malt
½	lb. (0.23 kg.)	German Munich malt
3.2	oz. (90 g.)	German Sauer (sour) malt
3	**lbs. (1.4 kg.)**	**Total grains**

10 HBU (284 MBU) German Hersbrucker Spalt hops (whole)—60 minutes (bittering)

½ oz. (14 g.) Czech Saaz hops (whole)—steep in finished boiled wort for 2 to 3 minutes (aroma)

¼ tsp. Irish moss

¾ c. corn sugar for priming in bottles. Use ⅓ cup corn sugar if priming a keg.

Wyeast 2278 Czech Pils yeast or other lager yeast producing a crisp finish, low malt character and good attenuation with no esters, sulfur or diacetyl.

A double-decoction mash is employed to mash the grains. If you are using alkaline or hard water, treat the water with an appropriate amount of lactic acid or refer to pages 120 to 122 of *The Home Brewer's Companion* for proceeding with a triple decoction with an acid rest.

Add 3 quarts (3 l.) of boiling water to the grain, raising the temperature to 122 degrees F (50 C). Then remove 1.5 quarts (1.4 l.) of the thickest mash and boil this in another vessel for 30 minutes. Stir this boiled decoction constantly. Maintain the remainder of the mash at 122 degrees F (50 C) while boiling the decocted portion. After boiling for 30 minutes, add the decoction back into the main mash vessel, raising the temperature to 150 degrees F (65.5). Maintain this temperature for one hour. Up to 3 quarts (8.6 l.) of boiling water may be added at any time during this period to maintain mashing temperature. After one hour remove one third of the mash (a fifty-fifty blend of thick mash and liquid) and boil about 20 minutes. Stir this second decoction constantly during the boil. When finished return the boiled decoction to the main mash vessel, ending starch conversion by having raised the temperature to about 167 degrees F (75 C). Then lauter and sparge with 1.5 gallons (5.7 l.) of 170-degree F (77 C) water. Collect about 2.5 gallons (9.5 l.) of runoff, add bittering hops and bring to a full and vigorous boil.

The total boil time will be 60 minutes. When 15 minutes remain, add flavor hops and Irish moss. After a total wort boil of 90 minutes (reducing the wort volume to just over 5 gallons), turn off the heat, then add aroma hops and let steep 2 to 5 minutes. Then separate or strain out and sparge hops, and direct the hot wort into a sanitized fermenter to which 2 gallons (7.6 l.) of cold water have been added. If necessary, add additional cold water to achieve a 5-gallon (19-l.) batch size. Chill the wort to 60 degrees F (15.5 C) and direct into a sanitized fermenter. Aerate the cooled wort well. Add an active yeast culture and ferment for 4 to 6 days in the primary at about 55 degrees F (13 C). Then transfer into a secondary fermenter, chill to 50 degrees F (10 C) to age for two more weeks, then lager for two to four more weeks at 40 degrees F (4.5 C).

When secondary aging is complete, prime with sugar, bottle or keg. Let condition at temperatures above 60 degrees F (15.5 C) until clear and carbonated, then store chilled.

Silver Cup Winner

Brooklyn Lager
Brooklyn Brewery
Brooklyn, New York, USA

Amber color with an orange hue. Remarkably hoppy with noble-type hop character such as with Hallertauer Mittelfrüh hops. Some caramel notes in aroma, but it is the gentle, soft aroma of European-grown hops that predominates, or so it seems. Munich malt or toasted-malt character also is evident. Malt aromatics are expressed as an accent and are not overdone. Medium-bodied mouth feel. Noble hop flavor is quickly followed by the gentle toasted character of Munich-type malts and light caramel malt. Hop floral character could easily be mistaken for malt. While this beer is not exceptionally bitter, hops provide balance and linger enough to refresh the palate. Extraordinarily clean with no diacetyl. Well done with malt, hops and careful lagering. Is there Saaz in this beer?

Estimated profile based on tasting
Color: 11–13 SRM (22–26 EBC)
Bittering Units: 28–32

Bronze Cup Winner

Bohemian Pilsner
Bohemian Brewery
Torrance, California, USA

Pale golden lager with a hint of orange. Sensual aroma of floral and herbal hops. Uniquely suggestive of lavender in the aroma with pale malt sweetness as a foundation. The aroma dances with foxes. Medium- to light-bodied with the grounded, gentle, soft sweetness of skillfully processed pale malts. Soft hop bitterness slowly emerges to crown the experience. Bitterness lingers gently in aftertaste and is a real attraction of this beer. Malt and hops are both present, yet don't interfere with each other. A great balance of opposites. Soft and sensual, but definitive in manner. Clean, well balanced and a great statement for this style of beer. Lacking some of the well-lagered full-malt character of the Gold Cup winner.

Estimated profile based on tasting
Color: 5 SRM (10 EBC)
Bittering Units: 32–35

CATEGORY 37:
EUROPEAN-STYLE PILSENER

Continental Pilseners are straw/golden in color and are well-attenuated beers. This medium-bodied beer is often brewed with rice, corn, wheat or other grain or sugar adjuncts. Fruity esters and diacetyl should not be perceived. There should be no chill haze.

Original Gravity (°Plato):	1.044–1.050 (11–12.5 °Plato)
Apparent Extract– Final Gravity (°Plato):	1.008–1.010 (2–2.5 °Plato)
Alcohol by weight (volume):	3.6–4.2% (4–5%)
Bitterness (IBU):	17–30
Color (SRM):	3–4 (7–10 EBC)

Gold Cup Winner

Dos Equis Special Lager
Cerveceria Cuauhtèmoc Moctezuma S.A. de C.V.
Ave. Alfonso Reyes 2202 Nte.
Apartado Postal 106
Monterrey, Nuevo Leon, Mexico 64442
Brewmaster: Adolfe Vallesteros
Established 1890
Production: 18.8 million bbl. (22 million hl.)

Cecilia Alvarez-Tabil

At the main brewery site in Monterrey, one immediately notes the monumental 60-meter-tall redbrick tower, built in 1905 as a cornerstone of the proud American tradition of Cerveceria Cuauhtèmoc Moctezuma. Carta Blanca was the first beer to be marketed in 1890 by the brewery. In 1908 King Alfonso XIII of Spain named Cuauhtèmoc the official provider of the Royal House. Since then the original brewery, merged in 1985 with Moctezuma brewery, has grown into an expansive international presence.

Celebrating 100 years in 1990, the Cuauhtèmoc Moctezuma brewery is adorned with beer awards. Recognized in 1893 for the purity and quality of the brand Carta Blanca at the International Exhibition in Chicago, the brewery went on to win a Gold Medal in 1900 at the Universal Exhibition of Paris. The Cuauhtèmoc brewery won another award in 1905 at the International Exhibition in Leija, Belgium, and in 1991 the beer Carta Blanca was recognized at the Monde Selection Prix in Paris. Most recently its Dos Equis Special Lager won the Gold in the 1996 World Beer Cup.

The brewery's other products, which are brewed and distributed worldwide, include Carta Blanca, Bohemia, Indio,

Tecate, Chihuahua, Superior, Dark Superior, Dark Dos Equis, Sol, Noche Buena, Tecate Light and Tecate Twist.

CHARACTER DESCRIPTION OF GOLD CUP–WINNING DOS EQUIS
SPECIAL LAGER

Light golden color. The beer is packaged in a green bottle and has a surprisingly pleasant, very subtle light-struck character (yes, I said "pleasant"—and mean it) that accents the hop character. First aromatic impression is reminiscent of the aroma of the Czech Pilsner Urquel with its Saaz hop aroma. The brewery uses a combination of Galena and Cluster hops only for bittering and does not claim to use Saaz in this beer. (It does use Saaz in Bohemia Lager.) Mouth feel is light in body and exceptionally clean. Though brewed with corn and rice, the overall character suggests an all-malt character with high drinkability. Hop bitterness is very low but perceptible and balanced with a light touch of sweet malt flavor and aroma. A beer that does not obfuscate. Simple, clean and refreshing with a foundation of quality beer character.

- **Recipe for 5 U.S. gallons (19 liters) Dos Equis Special Lager**
  Targets:
  Original Gravity: 1.060 (15)
  Final Gravity: 1.014 (3.5)
  Alcohol by volume: 6%
  Color: 14 SRM (28 EBC)
  Bittering Units: 60

ALL-GRAIN RECIPE AND PROCEDURE

4¾ lbs.	(2.2 kg.)	American 6-row pale malt
2½ lbs.	(1.1 kg.)	rice
¾ lb.	(0.34 kg.)	flaked corn
8 lbs.	**(3.6 kg.)**	**Total grains**

1.5   HBU   (43 MBU) American Cluster hops (pellets)—75 minutes (bittering)

2   HBU   (56 MBU) American Hallertauer hops (pellets)—75 minutes (bittering)

2   HBU   (56 MBU) Czech Saaz hops (pellets)—10 minutes (flavor)

¼   tsp.   Irish moss

¾   c.   corn sugar for priming in bottles. Use ⅓ cup corn sugar if priming a keg.

Wyeast 2007 Pilsen Lager yeast, which ferments dry with a malty finish. No sulfur, diacetyl or esters.

A rice-cooking regime along with a step infusion mash is employed to mash the grains. Crush and mill rice into small pieces. Add crushed rice to 1 gallon (3.8 l.) of water and boil for 20 minutes. Add one more gallon of water to cooked rice mash to achieve a temperature of 130 degrees F (54.5 C). Add flaked corn and malt; stabilize and hold the temperature at 122 degrees F (50 C) for 30 minutes. Add 4 quarts (3.8 l.) of boiling water, adding heat if necessary to bring temperature up to 150 degrees F (65.5 C). Hold for about 60 minutes.

After conversion, raise temperature to 167 degrees F (75 C), lauter and sparge with 4 gallons (15 l.) of 170-degree F (77 C) water. Collect about 6 gallons (23 l.) of runoff, add bittering hops and bring to a full and vigorous boil.

The total boil time will be 75 minutes. When 10 minutes remain, add flavor hops and Irish moss. After a total wort boil of 75 minutes (reducing the wort volume to just over 5 gallons), turn off the heat, then separate or strain out and sparge hops. Chill the wort to 60 degrees F (15.5 C) and direct into a sanitized fermenter. Aerate the cooled wort well. Add an active yeast culture and ferment for 4 to 6 days in the primary at about 55 degrees F (13 C). Then transfer into a secondary fermenter, chill to 50 degrees F (10 C) to age for two more weeks, then lager for two to four more weeks at 40 degrees F (4.5 C).

When secondary aging is complete, prime with sugar, bottle or keg. Let condition at temperatures above 60 degrees F (15.5 C) until clear and carbonated, then store chilled.

MALT-EXTRACT RECIPE AND PROCEDURE FOR DOS EQUIS SPECIAL LAGER

3	lbs. (1.4 kg.)	English extralight dried malt extract
2½	lbs. (1.1 kg.)	rice extract
0	**lb. (0 kg.)**	**Total grains**

2	HBU	(56 MBU) American Cluster hops (pellets)—60 minutes (bittering)
2	HBU	(56 MBU) American Hallertauer hops (pellets)—60 minutes (bittering)
2	HBU	(56 MBU) Czech Saaz hops (pellets)—10 minutes (flavor)

¼	tsp.	Irish moss
¾	c.	corn sugar for priming in bottles. Use ⅓ cup corn sugar if priming a keg.

Wyeast 2007 Pilsen Lager yeast, which ferments dry with a malty finish. No sulfur, diacetyl or esters.

Add the dried malt extract, rice extract and bittering hops to 2.5 gallons (9.5 l.) of water and bring to a full and vigorous boil.

The total boil time will be 60 minutes. When 10 minutes remain, add flavor hops and Irish moss. After a total wort boil of 60 minutes, turn off the heat, separate or strain out and sparge hops, and direct the hot wort into a sanitized fermenter to which 2 gallons (7.6 l.) of cold water have been added. If necessary, add additional cold water to achieve a 5-gallon (19-l.) batch size. Chill the wort to 60 degrees F (15.5 C). Aerate the cooled wort well. Add an active yeast culture and ferment for 4 to 6 days in the primary at about

55 degrees F (13 C). Then transfer into a secondary fermenter, chill to 50 degrees F (10 C) to age for two more weeks, then lager for two to four more weeks at 40 degrees F (4.5 C).

When secondary aging is complete, prime with sugar, bottle or keg. Let condition at temperatures above 60 degrees F (15.5 C) until clear and carbonated, then store chilled.

Silver Cup Winner

Michael Shea's Blonde Lager
HighFalls Brewing Co.
Rochester, New York, USA

Pale golden color. Floral hop and very sweet base malt aroma. Low bitterness level gives a first flavor impression. Very sweet flavor with little bitterness in aftertaste. Very clean, smooth and medium-bodied lager. Smoothness indicates a well-lagered character. Unobtrusive sweetness of pale malt (no specialty malts perceived) is memorable. Exceptionally clean with some bitterness lingering after the first four or five mouthfuls. Mildly flavored friendly drinking lager.

Brewery formulation uses a blend of 6- and 2-row pale malts and 22 percent corn with American Galena and Mt. Hood hops for bitterness and aroma.

Estimated profile based on tasting
Original Gravity: 1.046 (11.5) indicated by the brewery
Final Gravity: 1.016 (4.1) indicated by the brewery
Alcohol by volume: 3.7% indicated by the brewery
Color: 4–5 SRM (8–10 EBC) (4.2 indicated by the brewery)
Bittering Units: 18–22 (18 indicated by the brewery)

Bronze Cup Winner

Efes Pilsen Erciyas, Efes Extra Special Pilsener
Erciyas Biracilik Ve Malt San, A. S.
Istanbul, Turkey

Light golden color. Very good pale-malt sweetness in aroma and flavor helps define the style. Character of toasted malt flirts with aroma and flavor, but never becomes definitive. Medium-bodied light lager. Bitterness emerges while I'm indulging to clean the palate. Not a soft bitterness. Clean and well balanced with a good dose of distinctive and somewhat harsh bitterness. Lasting impression is of a full-flavored light Pilsener. Alcohol is warming, suggesting that this may be stronger than usual. Strong, full-flavored, medium-bodied, satisfying, Pilsener beer with clean higher-than-usual alcohol.

Brewery formulation uses 25 percent rice with Turkish and German hops for bittering and aroma.

Estimated profile based on tasting
Original Gravity: 1.046 (11.5) indicated by the brewery
Final Gravity: 1.005 (1.2) indicated by the brewery
Alcohol by volume: 5.2% as claimed by the brewery, though packaging states 6% w w
Color: 4–5 SRM (8–10 EBC) (6 EBC indicated by the brewery)
Bittering Units: 20–26 (21 indicated by the brewery)

CATEGORY 38:
EUROPEAN-STYLE LOW-ALCOHOL LAGER/GERMAN-STYLE LEICHT

Very light in body and color. Malt sweetness is perceived at low to medium levels while hop bitterness character is perceived at medium levels. These beers should be clean with no perceived fruity esters or diacetyl. Chill haze is not acceptable.

Original Gravity (°Plato):	1.026–1.032 (6.5–8 °Plato)
Apparent Extract– Final Gravity (°Plato)	1.006–1.010 (1.5–2.5 °Plato)
Alcohol by weight (volume):	2.0–2.6% (2.5–3.3%)
Bitterness (IBU):	17–28
Color (SRM):	2–4 (5–9 EBC)

Gold Cup Winner

None

Silver Cup Winner

Egger Leicht
Privatbrauerei Fritz Egger G.m.b.H.
St. Pölten, Austria

Pale straw color with a rich, dense head. Perfectly balanced bready malt character in aroma reveals a German decoction-mashing method, perfect for style. Notable noble hop aroma. Light in body, and a delightful full flavor despite its lightness. Bitterness is soft and associates well with a delicious malt character. The two are completely in balance, both orbiting in equilibrium around a central theme. The beer speaks for itself. One of the best light-flavored beers I've ever had. Clean, quenching, dry finish. Exceptionally clean. Perhaps judges perceived some oxidation (which I don't) and that is why it did not get a Gold. A style of beer not yet brewed in America. Decoction-mash character is a signature of this beer. Skillfully accomplished by a knowledgeable brewmaster.

Estimated profile based on tasting
Original Gravity: 1.032–1.034 (8–8.5) estimated from observation
Final Gravity: 1.007 (1.77) indicated by the brewery
Alcohol by volume: 3.5% as indicated on the label (3.67% indicated by the brewery)
Color: 3 SRM (6 EBC) (6.7 indicated by the brewery)
Bittering Units: 27–30 (25 indicated by the brewery)

Bronze Cup Winner

None

CATEGORY 39:
MÜNCHNER-STYLE HELLES

This beer has a relatively low bitterness. It is a medium-bodied, malt-emphasized beer; however, certain versions can approach a balance of hop character to maltiness. There should not be any caramel character. Color is light straw to golden. Fruity esters and diacetyl should not be perceived.

Original Gravity (°Plato):	1.044–1.050 (11–13 °Plato)
Apparent Extract–	
Final Gravity (°Plato):	1.008–1.012 (2–3 °Plato)
Alcohol by weight (volume):	3.8–4.4% (4.5–5.5%)
Bitterness (IBU):	18–25
Color (SRM):	3–5 (7–12 EBC)

Gold Cup Winner

Bow Valley Premium Lager
Bow Valley Brewing Company
109 Boulder Crescent
Canmore, Alberta, Canada T0L 0M0
Brewmaster: Gordon Demaniuk
Established 1995
Production: 2,700 bbl. (3,000 hl.)

Hugh Hancock

left to right, *Jim Ridley, Delivery and Sales; Gord Demaniuk, Brewmaster; Wendy MacMillan, Asst. Brewmaster; and Hugh Hancock, President*

Their first batch of beer was proudly begun. The malt was milled and mashed into the new vessels. Then at midnight, January 31, 1995, high in the Canadian Rocky Mountains, not far from Banff National Park, the mash tuns' plates collapsed, unable to support the weight of the mash. The next phase seemed to establish the spirit of success as friends, family and brewery workers scrambled for twenty hours. The first batch was finally piped to the fermenters. Strewn about the brewery, friends, workers, and family sacked out on bags of malt used as makeshift mattresses. The brewery's final word on this episode was, "We got better at it."

For centuries, brewing the first exciting batch of beer and establishing an appreciative market have been malt for the mill of brewery tales. Bow Valley, incorporated in 1993, is part of the world family of breweries keeping this tradition alive. It has had its share of adversity and tribulations. All small brewing companies do. We read of breweries established 200, 300, 500 years ago, and may attribute a certain measure of sacredness to their antiquity. Just as sacred are the family and spirit of friendship that play an important role in sustaining any brewery, new or old.

Originally built to produce at a 6,400-barrel (7,500-hl.) capacity, Bow Valley offers two original beers: Bow Valley Premium Lager and Bruno's Mountain Bock, both all-malt beers brewed according to the German purity law.

CHARACTER DESCRIPTION OF GOLD CUP–WINNING BOW VALLEY PREMIUM LAGER

Deep golden color with no amber hue. Clear with no chill haze. Mild and pleasant Hallertauer-type hop aroma. The key descriptor for hop aroma is *mild*, though it is evident to the enthusiast. Malt aroma is simple yet not sweet. Fermentation is exceptionally clean. Bitterness is appropriately mild and not particularly notable. First taste impression indicates high carbonation, low body, mild flavor. Not highly hopped. Malt sweetness lacking in assertiveness. This style of beer is meant to be enjoyed in *mas* liter mugs with great German-style swigs and swallows. With its gentleness and underlying hop and all-malt flavor, the full effect and mouth feel of a Munich Helles is achieved. A winner, but relatively dry and light-bodied for style.

• **Recipe for 5 U.S. gallons (19 liters) Bow Valley Premium Lager**
Targets:
Original Gravity: 1.046 (11.5)
Final Gravity: 1.010 (2.5)
Alcohol by volume: 5%
Color: 6 SRM (12 EBC)
Bittering Units: 17

ALL-GRAIN RECIPE AND PROCEDURE

7 ¾ lbs. (3.5 kg.)   American 2-row Klages/Harrington pale malt

½ lb.   (0.23 kg.)   American Munich malt

2 ½ oz. (70 g.)   English Carastan malt—40 Lovibond

8 ½ lbs. (3.9 kg.)   Total grains

2.5 HBU (71 MBU) American Nugget hops (pellets)—75 minutes (bittering)

1   HBU (28 MBU) American Mt. Hood hops (pellets)—30 minutes (flavor)

1   HBU (28 MBU) Hersbrucker Hallertauer hops (pellets)—30 minutes (flavor)

1   HBU (28 MBU) Hersbrucker Hallertauer hops (pellets)—15 minutes (second flavor)

1   HBU (28 MBU) Czech Saaz hops (pellets)—15 minutes (second flavor)

¼   tsp. Irish moss

¾   c.   corn sugar for priming in bottles. Use ⅓ cup corn sugar if priming a keg.

Wyeast 2308 Munich Lager yeast or Wyeast 2206 Bavarian Lager yeast; both have a smooth, clean finish with a malt accent.

A single-step infusion mash is employed to mash the grains. Add 8.5 quarts (6.7 l.) of 172-degree F (78 C) water to the crushed grain, stir, stabilize and hold the temperature at 155 degrees F (68 C) for 60 minutes.

After conversion, raise temperature to 167 degrees F (75 C), lauter and sparge with 4 gallons (15 l.) of 170-degree F (77 C) water. Collect about 5.5 gallons (21 l.) of runoff, add bittering hops and bring to a full and vigorous boil.

The total boil time will be 75 minutes. When 30 minutes remain, add flavor hops. When 15 minutes remain, add second flavor hops and Irish moss. After a total wort boil of 75 minutes (reducing the wort volume to just over 5 gallons), turn off the heat, then separate or strain out and sparge hops. Chill the wort to 60 degrees F (15.5 C) and direct into a sanitized fermenter. Aerate the cooled wort well. Add an active yeast culture and ferment for 4 to 6 days in the primary at about 55 degrees F (13 C). Then transfer into a secondary fermenter, chill to 50 degrees F (10 C) to age for two more weeks, then lager for two to four more weeks at 40 degrees F (4.5 C).

When secondary aging is complete, prime with sugar, bottle or keg. Let condition at temperatures above 60 degrees F (15.5 C) until clear and carbonated, then store chilled.

MASH-EXTRACT RECIPE AND PROCEDURE FOR BOW VALLEY PREMIUM LAGER

3 ¼ lbs.	(1.6 kg.)	English extralight dried malt extract
2 ½ lbs.	(1.1 kg.)	American 2-row Kalges/Harrington pale malt
½ lb.	(0.23 kg.)	American Munich malt
2 ½ oz.	(70 g.)	English Carastan malt—40 Lovibond
3 lbs.	**(1.4 kg.)**	**Total grains**

3	HBU	(85 MBU) American Nugget hops (pellets)—60 minutes (bittering)
1	HBU	(28 MBU) American Mt. Hood hops (pellets)—30 (flavor)
1	HBU	(28 MBU) Hersbrucker Hallertauer hops (pellets)—30 minutes (flavor)
1	HBU	(28 MBU) Hersbrucker Hallertauer hops (pellets)—15 minutes (second flavor)
1	HBU	(28 MBU) Czech Saaz hops (pellets)—15 minutes (second flavor)

¼	tsp.	Irish moss
¾	c.	corn sugar for priming in bottles. Use ⅓ cup corn sugar if priming a keg.

Wyeast 2308 Munich Lager yeast or Wyeast 2206 Bavarian

Lager yeast; both have a smooth, clean finish with a malt accent.

A single-step infusion mash is employed to mash the grains. Add 3 quarts (3 l.) of 172-degree F (78 C) water to the crushed grain, stir, stabilize and hold the temperature at 155 degrees F (68 C) for 60 minutes.

After conversion, raise temperature to 167 degrees F (75 C), lauter and sparge with 2 gallons (7.6 l.) of 170-degree F (77 C) water. Collect about 2.5 gallons (9.5 l.) of runoff. Add malt extract and bittering hops and bring to a full and vigorous boil.

The total boil time will be 60 minutes. When 30 minutes remain, add flavor hops. When 15 minutes remain, add second flavor hops and Irish moss. After a total wort boil of 60 minutes, turn off the heat. Then separate or strain out and sparge hops, and direct the hot wort into a sanitized fermenter to which 2 gallons (7.6 l.) of cold water have been added. If necessary, add additional cold water to achieve a 5-gallon (19-l.) batch size. Chill the wort to 60 degrees F (15.5 C) and direct into a sanitized fermenter. Aerate the cooled wort well. Add an active yeast culture and ferment for 4 to 6 days in the primary at about 55 degrees F (13 C). Then transfer into a secondary fermenter, chill to 50 degrees F (10 C) to age for two more weeks, then lager for two to four more weeks at 40 degrees F (4.5 C).

When secondary aging is complete, prime with sugar, bottle or keg. Let condition at temperatures above 60 degrees F (15.5 C) until clear and carbonated, then store chilled.

Silver Cup Winner

Granville Island Lager
Granville Island Brewing Co.
Vancouver, British Columbia, Canada

Light golden color. Aroma particularly rich and malty, with floral sweetness of hops providing support. Medium body with a clean, light, dry finish. Hop flavor is soft and suggestive rather than assertive. Of particular note is the soft character of the bitterness and the full malt flavor, accomplished with a skillful lightness, enhancing drinkability. Remarkably clean, with no unusual fermented character other than what the malt and hops bring to the beer. Smooth well-aged lager character provides assurance. A beer that can be enjoyed quite cold, but is best at traditional Münchner-style serving temperatures, which provide a big difference in the experience.

Estimated profile based on tasting
Alcohol by volume: 5% as stated on the label
Color: 4–5 SRM (8–10 EBC)
Bittering Units: 22–26

Bronze Cup Winner

Scrimshaw (Pilsener Style) Beer
North Coast Brewing Co., Inc.
Fort Bragg, California, USA

Pale golden color. Floral hop aroma suggests American-grown nobles. Pleasant and rich pale-malt base provides fundamental aroma. Light-bodied with a tendency toward becoming medium-bodied. Clean, soft bitterness with a dry, refreshing finish. No faults perceived. Hop flavor comes through nicely. Aftertaste is memorably bitter but not obtrusive. Excellent balance for Pils. Smooth, well-lagered character. No fruitiness or fermentation faults perceived. Though a bit dry and light for style, rather excellent portrayal of a German-style Pilsener.

Estimated profile based on tasting
Color: 4–5 SRM (8–10 EBC)
Bittering Units: 30–35

CATEGORY 40: DORTMUNDER/ EUROPEAN-STYLE EXPORT

Both starting gravity and bitterness are somewhat higher than a Münchner Helles. Bitterness is medium, hop flavor and aroma are perceptible but low. The color of this style may be slightly darker than Helles, and the body is fuller but still medium-bodied. Fruity esters, chill haze and diacetyl should not be perceived.

Original Gravity (°Plato):	1.048–1.056 (12–14 °Plato)
Apparent Extract– Final Gravity (°Plato):	1.010–1.014 (2.5–3.5 °Plato)
Alcohol by weight (volume):	4–4.8% (5–6%)
Bitterness (IBU):	23–29
Color (SRM):	3–5 (8–13 EBC)

Gold Cup Winner

Stoudt's Export Gold
Stoudt's Brewing Co.
P.O. Box 880
Route 272
Adamstown, Pennsylvania, USA 19501
Brewmaster: Carol Stoudt
Established 1987
Production: 7,000 bbl. (8,200 hl.)

Located in the hills of Pennsylvania Dutch Country, not far from the towns of Intercourse and Blue Balls, one of the premier microbreweries in the United States brews twelve different styles of beer, available year-round. Founded in 1987 by Carol Stoudt, one of only a handful of woman brewmasters in North America, Stoudt's brewery makes the extra effort to brew to tradition, whether it's a decoction German lager, Belgian Trippel, English bitter or American pale ale. Attached to the brewery is a large *biergarten* open daily during the warmer months of the year. It is the setting for eastern Pennsylvania's Annual Summer Bier Festival. Husband Ed Stoudt owns and operates the Black Angus Restaurant, adjacent to the brewery.

Carol A. Stoudt

Carol Stoudt, Brewer

For Carol and Ed, it all began on a bus trip in 1986, on an Institute for Brewing Studies–sponsored tour of the microbreweries of the Northwest. (There were perhaps a total of twelve in Oregon, Washington and British Columbia.) Inspired by what they saw, they opened the first microbrewery in Pennsylvania since the decade of American Prohibition began in 1923.

Brewmaster Carol Stoudt has won seventeen Gold, Silver and Bronze medals at the Great American Beer Festival since the brewery opened. Stoudt's Export Gold had already won five separate Gold, Silver and Bronze medals in the GABF before coming out on top at the World Beer Cup. Other products brewed by Carol at the Stoudt Brewery include Stoudt's Amber, Scarlet Lady Ale E.S.B., Fat Dog Stout, Abbey Double, Abbey Triple, Stout's (Oktober) Fest, Stoudt's I.P.A., Winter Ale, Stoudt's Bock, Stoudt's Doublebock, Stoudt's Pilsner, Stoudt's Mai Bock, Stoudt's Weizen and Stoudt's Pale Ale.

CHARACTER DESCRIPTION OF GOLD CUP–WINNING STOUDT'S EXPORT GOLD

Light golden amber in color. Malty, bready, toasted-malt

aroma. Breadiness can be attributed to a decoction mash. When the beer is served at colder temperatures, hop aroma is not perceived. As beer temperature increases to German-style beer-drinking temperatures, hop aroma is released, which has a soft, sweet, floral hop character such as that produced by American Tettnanger. American noble hop aroma. Tettnanger aroma very subtle; Tettnanger is in fact used with a combination of other American-grown nobles. Munich/Vienna malt character with hints of caramel dominates, effectively suppressing noble hop character. Flavor suggests the use of a very small amount of caramel, and indeed this is the case in formulation. A well-rounded malt character dominates in the flavor impression. Balanced soft bitterness does not overwhelm. Body is medium in character. Softness of malt flavor and toasted biscuitlike malt aroma permeate as notable themes, but do not assault. Great drinkability. Absolutely no diacetyl, sulfur or fruity esters perceived.

- **Recipe for 5 U.S. gallons (19 liters) Stoudt's Export Gold**
  Targets:
  Original Gravity: 1.050 (12.5)
  Final Gravity: 1.012 (2.5)
  Alcohol by volume: 5%
  Color: 8 SRM (16 EBC)
  Bittering Units: 25

ALL-GRAIN RECIPE AND PROCEDURE

8	lbs.	(3.6 kg.)	American 2-row pale malt
½	lb.	(0.23 kg.)	American Munich malt
¼	lb.	(114 g.)	American Cara-Pils malt
¼	lb.	(114 g.)	American Victory or other aromatic malt
2½	oz.	(70 g.)	American caramel malt—40 Lovibond
9.2	**lbs.**	**(4.2 kg.)**	**Total grains**

2.5 HBU (71 MBU) American Cluster hops (pellets)—90 minutes (bittering)

1.5 HBU (43 MBU) American Hallertauer hops (pellets)—90 minutes (bittering)

1 HBU (28 MBU) American Hallertauer hops (pellets)—30 minutes (flavor)

1 HBU (28 MBU) American Tettnanger hops (pellets)—30 minutes (flavor)

¼ oz. (7 g.) American Saaz hops (pellets)—steep in finished boiled wort for 2 to 3 minutes (aroma)

½ oz. (14 g.) American Tettnanger hops (pellets)—steep in finished boiled wort for 2 to 3 minutes (aroma)

¼ tsp. Irish moss

¾ c. corn sugar for priming in bottles. Use ⅓ cup corn sugar if priming a keg.

Wyeast 2308 Munich Lager yeast or Wyeast 2206 Bavarian Lager yeast; smooth, clean finish with malt accent.

A double-decoction mash is employed to mash the grains. If you are using alkaline or hard water, treat the water with an appropriate amount of lactic acid or refer to pages 120 to 122 of *The Home Brewer's Companion* (Avon Books, 1994) for proceeding with a triple decoction with an acid rest.

Add 9 quarts (8.6 l.) of boiling water to the grain, raising the temperature to 122 degrees F (50 C). Then remove 4.5 quarts (4.3 l.) of the thickest mash and boil this in another vessel for 30 minutes. Stir this boiled decoction constantly. Maintain the remainder of the mash at 122 degrees F (50 C) while boiling the decocted portion. After boiling for 30 minutes, add the decoction back into the main mash vessel, raising the temperature to 150 degrees F (65.5). Maintain this temperature for one hour. Up to 9 quarts (8.6 l.) of boiling water may be added at any time during this period to maintain mashing temperature. After one hour remove one third of the mash (a fifty-fifty blend of thick mash and liquid) and boil about 20 minutes. Stir this second decoction constantly during the boil. When finished, return the boiled decoction to the main mash vessel, ending starch conversion by having raised the temperature to about 167 degrees F (75 C). Then lauter and sparge with 4 gallons (15 l.) of 170-degree F (77 C) water. Collect about 6 gallons (23 l.) of runoff, add bittering hops and bring to a full and vigorous boil.

The total boil time will be 90 minutes. When 30 minutes remain, add flavor hops. When 10 minutes remain, add Irish moss. After a total wort boil of 90 minutes (reducing the wort volume to just over 5 gallons), turn off the heat, add aroma hops and let steep for 2 to 3 minutes, then separate or strain out and sparge hops. Chill the wort to 60 degrees F (15.5 C) and direct into a sanitized fermenter. Aerate the cooled wort well. Add an active yeast culture and ferment for 4 to 6 days in the primary at about 55 degrees F (13 C). Then transfer into a secondary fermenter, chill to 50 degrees F (10 C) to age for two more weeks, then lager for two to four more weeks at 40 degrees F (4.5 C).

When secondary aging is complete, prime with sugar, bottle or keg. Let condition at temperatures above 60 degrees F (15.5 C) until clear and carbonated, then store chilled.

MASH-EXTRACT RECIPE AND PROCEDURE FOR STOUDT'S EXPORT GOLD

3 ¾	lbs.	(1.7 kg.)	English extralight dried malt extract
2	lbs.	(0.9 kg.)	American 2-row pale malt
½	lb.	(0.23 kg.)	American Munich malt
¼	lb.	(114 g.)	American Cara-Pils malt
¼	lb.	(114 g.)	American Victory or other aromatic malt
2 ½	oz.	(70 g.)	American caramel malt—40 Lovibond
3.2	**lbs.**	**(1.5 kg.)**	**Total grains**

2.5	HBU	(71 MBU) American Cluster hops (pellets)—60 minutes (bittering)
2.5	HBU	(71 MBU) American Hallertauer hops (pellets)—60 minutes (bittering)
1	HBU	(28 MBU) American Hallertauer hops (pellets)—30 minutes (flavor)
2	HBU	(56 MBU) American Tettnanger hops (pellets)—30 minutes (flavor)
¼	oz.	(7 g.) American Saaz hops (pellets)—steep in finished boiled wort for 2 to 3 minutes (aroma)
½	oz.	(14 g.) American Tettnanger hops (pellets)—steep in finished boiled wort for 2 to 3 minutes (aroma)
¼	tsp.	Irish moss
¾	c.	corn sugar for priming in bottles. Use ⅓ cup corn sugar if priming a keg.

Wyeast 2308 Munich Lager yeast or Wyeast 2206 Bavarian Lager yeast; smooth, clean finish with malt accent.

A double-decoction mash is employed to mash the grains. If you are using alkaline or hard water, treat the water with an appropriate amount of lactic acid or refer to pages 120 to 122 of *The Home Brewer's Companion* for proceeding with a triple decoction with an acid rest.

Add 3 quarts (3 l.) of boiling water to the grain, raising the temperature to 122 degrees F (50 C). Then remove 1.5 quarts (1.4 l.) of the thickest mash and boil this in another vessel for 30 minutes. Stir this boiled decoction constantly. Maintain the remainder of the mash at 122 degrees F (50 C) while boiling the decocted portion. After boiling for 30 minutes, add the decoction back into the main mash vessel, raising the temperature to 150 degrees F (65.5). Maintain this temperature for one hour. Up to 3 quarts (8.6 l.) of boiling water may be added at any time during this period to maintain mashing temperature. After one hour remove one third of the mash (a fifty-fifty blend of thick mash and liquid) and boil about 20 minutes. Stir this second decoction constantly during the boil. When finished, return the boiled decoction to the main mash vessel, ending starch conversion by having raised the temperature to about 167 degrees F (75 C). Then lauter and sparge with 1.5 gallons (5.7 l.) of 170-degree F (77 C) water. Collect about 2.5 gallons (9.5 l.) of runoff, add bittering hops and bring to a full and vigorous boil.

The total boil time will be 60 minutes. When 30 minutes remain, add flavor hops. When 10 minutes remain, add Irish moss. After a total wort boil of 60 minutes (reducing the wort volume to just over 5 gallons), turn off the heat, then add aroma hops and let steep 2 to 5 minutes. Then separate or strain out and sparge hops, and direct the hot wort into a sanitized fermenter to which 2 gallons (7.6 l.) of cold water have been added. If necessary, add additional cold water to achieve a 5-gallon (19-l.) batch size. Chill the wort to 60 degrees F (15.5 C) and direct into a sanitized fermenter. Aerate the cooled wort well. Add an active yeast culture and ferment for 4 to 6 days in the primary at about 55 degrees F (13 C). Then transfer into a secondary fermenter, chill to 50 degrees F (10 C) to age for two more weeks, then lager for two to four more weeks at 40 degrees F (4.5 C).

When secondary aging is complete, prime with sugar, bottle or keg. Let condition at temperatures above 60 degrees F (15.5 C) until clear and carbonated, then store chilled.

Silver Cup Winner

Berghoff Original Lager Beer
Joseph Huber Brewing Co.
Monroe, Wisconsin, USA

Rich golden color with no hints of red or orange. Golden bubbles emerge on a summer's late afternoon as I'm noting the foam cling on the glass, a sign of quality, care and character. First aromatic impressions offer a sweet pale-malt character with a pleasantly low level of DMS (a sweet corn-like aroma). There is no hint of any caramel. A fully carbonated beer, yet Berghoff doesn't explode on your palate. The full malt flavor supports a pleasant backdrop of complementing hop bitterness without imposition. Though this is not a bitter beer, there is enough bitterness to provide enticement and drinkability. Wonderful example of a light-bodied to medium-bodied beer with substance. Clean with no complex fermented character. A beer that has American pride.

Brewery formulation uses American Cara-Pils and caramel malts and corn syrup, with CO_2 hop extract, American Galena and Tettnanger hops for bitterness, flavor and aroma.

Estimated profile based on tasting
Original Gravity: 1.052 (13.1) indicated by the brewery
Final Gravity: 1.020 (5) indicated by the brewery
Alcohol by volume: 5% indicated by the brewery
Color: 4–5 SRM (8–10 EBC) (5 indicated by the brewery)
Bittering Units: 22–24 (21 indicated by the brewery)

Bronze Cup Winner

BCC (Business Centre Club) ("Extra Premium")
LECH Browary Wielkopolski SA
Poznan, Poland

Golden yellow color. Clean beer with slight fruity-ester and pleasant floral hop aroma. Slight sweetness associated with hops. Flavor is regal and clean with a slight hop bitterness that is in balance with malt character of beer. Body is more light than medium. Remarkably clean aftertaste with a lingering malt and soft hop bitterness, the hops providing a refreshing afterbite. Extremely high drinkability and dry finish. First impression suggest malt and hops accenting appropriately for style, but with evolving second thoughts, hops and alcohol may be a bit on the low side for style.

Estimated profile based on tasting
Color: 4–5 SRM (8–10 EBC)
Bittering Units: 23–27

CATEGORY 41: VIENNA-STYLE LAGER

Beers in this subcategory are reddish brown or copper-colored. They have light to medium body. They are characterized by malty aroma, slight malt sweetness and clean hop bitterness. Noble hop aromas and flavor should be low to medium. Fruity esters, diacetyl and chill haze should not be perceived.

Original Gravity (°Plato):	1.048–1.056 (12–14 °Plato)
Apparent Extract– Final Gravity (°Plato):	1.012–1.018 (3–4.5 °Plato)
Alcohol by weight (volume):	3.8–4.3% (4.8–5.4%)
Bitterness (IBU):	22–28
Color SRM (EBC):	8–12 (16–30 EBC)

Gold Cup Winner

Leinenkugel's Red Lager
Jacob Leinenkugel Brewing Company
1–3 Jefferson Ave.
Chippewa Falls, Wisconsin, USA 54729
Brewmaster: John Buhrow
Established 1867
Production: 325,000 bbl. (380,000 hl.)

A dozen women and 2,500 loggers and sawmill workers lived in Chippewa Falls when Jacob Leinenkugel built his first brewery there in 1867. The workdays were long and the weather harsh. You've got to believe that homebrew was part of the in-

JACOB LEINENKUGEL BREWING COMPANY

spiration for Jacob, son of a Bavarian brewmaster. Up went the town's first brewery, all 1,200 square feet of it, and soon thereafter, the minds of the workers were off the weather, work and the lack of other amenities. The brewery introduced its bock beer in 1888 and expanded in 1890. It survived the American Prohibition producing Leinie's soda water. It wasn't until 1970 that the brewery distributed its beers outside of Wisconsin by quenching the thirst of neighboring Minnesota. Since then, five generations have continually called the shots for the company. The Miller Brewing Company entered into a joint venture with Leinenkugel in 1988. To many that means "bought," but much to the credit of both breweries, Leinenkugel has maintained much of its independent nature and is brewing a far greater variety of beers than it would have without its parent company.

Chippewa Falls, just 100 miles east of Minneapolis, has always been part of Wisconsin's romantic North Woods mystique. In the minds of millions of Americans, Leinenkugel was a family-run, hometown brewery. Anyone who knew beer in the Midwest knew of the legendary Leinenkugel beer. It has maintained its mystique and now more than ever finds itself in the midst of the specialty beer market revolution, brewing such specialties as Leinenkugel's Honey Weiss, Northwoods Lager, Auburn Ale, Genuine Bock, Big Butt Doppelbock, Berry Weiss, Autumn Gold, Winter Lager and Red Lager. Of course, it continues to brew the beer that claimed the thirst of midwesterners through many years, Leinenkugel's Original.

Awards are not new for Leinenkugel. In 1993 it won a Gold Medal in the Premium Lager category at the Great

American Beer festival. In 1987 Leinenkugel's won a Gold Medal in the Light Lager category at the GABF.

CHARACTER DESCRIPTION OF GOLD CUP–WINNING LEINENKUGEL'S RED LAGER

Evocatively deep orange-amber lager. Caramel and Munich malt character form first aromatic impressions. Subtle toasted-malt aroma remains throughout the experience along with a sweet floral hop aroma. Medium to low body with an exceptionally clean, dry finish. Perceived bitterness is at threshold, but due to high malt character, bitterness level is deceivingly high in analysis. A hint of dark caramel malt with a coffeelike roast note comes through without any hint of chocolate.

• **Recipe for 5 U.S. gallons (19 liters) Leinenkugel's Red Lager**
Targets:
Original Gravity: 1.048 (12)
Final Gravity: 1.014 (3.5)
Alcohol by volume: 4.8%
Color: 14 SRM (28 EBC)
Bittering Units: 18

ALL-GRAIN RECIPE AND PROCEDURE

2¾	lbs.	(1.2 kg.)	American 2-row Harrington pale malt
2¾	lbs.	(1.2 kg.)	American 6-row pale malt
1¾	lbs.	(0.8 kg.)	flaked corn
1½	lbs.	(0.69 kg.)	American caramel malt—50 Lovibond
¼	lb.	(114 g.)	American Cara-Pils malt
2½	oz.	(70 g.)	American Victory or other aromatic malt
9.2	**lbs.**	**(4.2 kg.)**	**Total grains**

3   HBU (85 MBU) American Clusters hops (pellets)—
            75 minutes (bittering)
3   HBU (85 MBU) American Mt. Hood hops (pellets)—
            20 minutes (flavor)

¼   tsp.   Irish moss
¾   c.   corn sugar for priming in bottles. Use ⅓ cup
             corn sugar if priming a keg.
Wyeast 2007 Pilsen Lager yeast can be recommended.

A step infusion mash is employed to mash the grains.
Add 9 quarts (8.5 l.) of 130-degree F (54.5 C) water to the
crushed grain, stir, stabilize and hold the temperature at
122 degrees F (50 C) for 30 minutes. Add 4.5 quarts (4.3 l.)
of boiling water, adding heat if necessary to bring temperature up to 150 degrees F (65.5 C). Hold for about 60
minutes.

After conversion, raise temperature to 167 degrees F (75
C), lauter and sparge with 4 gallons (15 l.) of 170-degree F
(77 C) water. Collect about 6 gallons (23 l.) of runoff, add
bittering hops and bring to a full and vigorous boil.

The total boil time will be 75 minutes. When 20 minutes
remain, add flavor hops. When 10 minutes remain, add Irish
moss. After a total wort boil of 75 minutes (reducing the
wort volume to just over 5 gallons), turn off the heat, then
separate or strain out and sparge hops. Chill the wort to 65
degrees F (18 C) and direct into a sanitized fermenter. Aerate the cooled wort well. Add an active yeast culture and
ferment for 4 to 6 days in the primary at 55 degrees F (15
C). Then transfer into a secondary fermenter, chill to 50
degrees F (10 C) to age for two more weeks, then lager for
two to four more weeks at 40 degrees F (4.5 C).

When secondary aging is complete, prime with sugar, bottle or keg. Let condition at temperatures above 60 degrees
F (15.5 C) until clear and carbonated, then store chilled.

MASH-EXTRACT RECIPE AND PROCEDURE FOR LEINENKUGEL'S RED LAGER

2½	lbs.	(1.1 kg.)	English light dried malt extract
2	lbs.	(0.9 kg.)	American 6-row pale malt
1	lb.	(0.45 kg.)	flaked corn
1½	lbs.	(0.69 kg.)	American caramel malt—50 Lovibond
¼	lb.	(114 g.)	American Cara-Pils malt
2½	oz.	(70 g.)	American Victory or other aromatic malt

4.95 lbs. (2.2 kg.)   Total grains

3.5 HBU (99 MBU) American Clusters hops (pellets)—75 minutes (bittering)

4 HBU (113 MBU) American Mt. Hood hops (pellets)—20 minutes (flavor)

¼ tsp. Irish moss

¾ c. corn sugar for priming in bottles. Use ⅓ cup corn sugar if priming a keg.

Wyeast 2007 Pilsen Lager yeast can be recommended.

A step infusion mash is employed to mash the grains. Add 5 quarts (4.8 l.) of 130-degree F (54.5 C) water to the crushed grain and flaked corn, stir, stabilize and hold the temperature at 122 degrees F (50 C) for 30 minutes. Add 2.5 quarts (2.4 l.) of boiling water. Add heat to bring temperature up to 150 degrees F (65.5 C). Hold for about 60 minutes.

After conversion, raise temperature to 167 degrees F (75 C), lauter and sparge with 2.5 gallons (9.5 l.) of 170-degree F (77 C) water. Collect about 3 gallons (11.5 l.) of runoff. Add malt extract and bittering hops and bring to a full and vigorous boil.

The total boil time will be 75 minutes. When 20 minutes remain, add flavor hops. When 10 minutes remain, add Irish moss. After a total wort boil of 75 minutes, turn off the heat, then separate or strain out and sparge hops, and direct the hot wort into a sanitized fermenter to which 2 gallons (7.6 l.) of cold water have been added. If necessary, add additional cold water to achieve a 5-gallon (19-l.) batch size. Chill the wort to 65 degrees F (68.5 C). Aerate the cooled wort well. Add an active yeast culture and ferment for 4 to 6 days in the primary at 55 degrees F (15 C). Then transfer into a secondary fermenter, chill to 50 degrees F (10 C) to age for two more weeks, then lager for two to four more weeks at 40 degrees F (4.5 C).

When secondary aging is complete, prime with sugar, bottle or keg. Let condition at temperatures above 60 degrees F (15.5 C) until clear and carbonated, then store chilled.

Silver Cup Winner

Dos Equis Amber Lager
Cerveceria Cuauhtèmoc
Monterrey, Mexico

Reddish orange amber lager. The caramel malt aroma is so mild I could be mistaken about it. Perhaps it is a mildly toasted Vienna or Munich malt rather than caramel, but probably it's a mixture of the two since its color belies a deepness not attributable to Munich or Vienna malts. First flavor impression suggests some corn-grainy character, adding to the complexity of the beer. Some subtle hop flavor in the sense of soft, mild herbal-wintergreen. Refreshing and suggestive light body evolves to a medium-bodied mouth feel, playing games with your sense of touch. Mild, clean, light caramel malt glimmers, but it's mostly pale malt offering sweetness. Aftertaste presents a pleasantly mild, refreshing bitterness that slowly evolves on the palate. Very clean finish.

Estimated profile based on tasting
Color: 11–13 SRM (22–26 EBC)
Bittering Units: 17–21

Bronze Cup Winner

Blue Ridge Amber Lager
Frederick Brewing Co.
Frederick, Maryland, USA

Orange-hued amber lager. Sweet toasted malt predominates in the aroma, with no hop aroma apparent. Flavor portrays an excellent balance between bitter and sweet; soft bitterness of hops is associated with light biscuit/toasted-malt sweetness, with neither one excessively accented, yet both in the spotlight. Medium-bodied mouth feel enhances richness of flavors. A beer relying on clean malt and hop flavors, both complex and endearing. Aftertaste lingers, reminiscing primarily about hops, both in flavor and bitterness. Hop bitterness offers a pleasurable sensation, skillfully balanced with body and full flavor of malts without being too heavy. There is some indication of a small amount of caramel malt, but principal character likely portrayed with a complex combination of toasted malts. Great and memorable beer.

Estimated profile based on tasting
Color: 13–14 SRM (26–28 EBC)
Bittering Units: 33–37

CATEGORY 42:
GERMAN-STYLE MÄRZEN/
OKTOBERFEST

Märzens are characterized by a medium body and broad range of color. Oktoberfests can range from golden to reddish brown. Sweet maltiness should dominate slightly over a clean hop bitterness. Male character should be of a toasted character rather than strongly caramel. Hop aroma and flavor should be low but notable. Fruity esters are minimal, if perceived at all. Diacetyl and chill haze should not be perceived.

Original Gravity (°Plato):	1.050–1.056 (12.5–14 °Plato)
Apparent Extract– Final Gravity (°Plato):	1.012–1.0120 (3–5 °Plato)
Alcohol by weight (volume):	4–47% (5.3–5.9%)
Bitterness (IBU):	18–25
Color (SRM):	4–15 (10–35 EBC)

Gold Cup Winner

None

Silver Cup Winner

Paulaner Oktoberfest
Paulaner Brewery
Munich, Germany

Rich, deep amber color with orange hues defines the quintessential color of the style. Aroma portrays an earthy bass note of pure maltiness with toasted candied character. Only the expert can sift through the complexities of the malt expression to find the subtly hidden hop aroma. Medium-to full-bodied mouth feel. A clean but light toasted-malt flavor reminiscent of freshly baked and toasted bread combines with hop flavor, aroma and bitterness. Overall the malty finish dominates. It is joined with just enough bitterness to further promote the malt expression, not neutralize it. Unlike underhopped light lagers, the invisible effect of adequate hops does not leave an acidic aftertaste on the palate. The burps are wonderful to taste and feel—an added bonus. Are we having fun? Can't help it with this brew. Clean aftertaste with the tangle of a little extra alcohol helping accent complexity and fruitiness. A classic.

Estimated profile based on tasting
Color: 11–13 SRM (22–26 EBC)
Bittering Units: 24–28

Bronze Cup Winner

Brasal Special Amber Lager
Brasserie Brasal Brewery
Lasalle, Québec, Canada

Rich, alluring copper-amber color. An exciting fullness and roundness suggested in aroma as wonderfully sweet Munich-type malt aroma emerges without an extreme toasted or biscuitlike character. Very much in character of style with a clean, gently toasted malt character integrated in flavor and aroma. The special malt flavor of an Oktoberfest-style lager definitively explodes on the palate with a follow-up soft hop bitterness immediately balancing the sweetness of malt. A clean finish due to the skillful balance between exceptional malt character and supporting hop bitterness. The malt character has a quiet suggestion of caramel toffee. Smooth and well lagered, this is a beer that has been shown patience and understanding. Aftertaste portrays a clean bitterness with memories of malt fading to promote another sip. This is a medium-bodied, noble all-malt beer of German character brewed in the French-speaking province of Canada.

Brewery formulation uses French Munich and caramel malts with American and German Perle, Spalter, Hallertauer and Tettnanger hops for bitterness, flavor and aroma.

Estimated profile based on tasting
Original Gravity: 1.060 (15) indicated by the brewery
Final Gravity: 1.014 (3.6) indicated by the brewery
Alcohol by volume: 6.1% indicated by the brewery
Color: 11–13 SRM (22–26 EBC) (24.5 EBC indicated by the brewery)
Bittering Units: 31–35 (22.5 indicated by the brewery)

CATEGORY 43:
EUROPEAN-STYLE DARK LAGER

A. SUBCATEGORY: EUROPEAN-STYLE DARK MÜNCHNER/DUNKEL

These beers have a pronounced malty aroma and flavor that dominate the clean, crisp moderate hop bitterness. A classic Münchner/Dunkel should have a chocolatelike, roast-malt, breadlike aroma that comes from the use of Munich dark malt. Chocolate or roast malts can be used, but the percentage used should be minimal. Noble hop flavor and aroma should be low but perceptible. Diacetyl is acceptable at very low levels. Fruity esters and chill haze should not be perceived.

Original Gravity (°Plato):	1.052–1.056 (13–14 °Plato)
Apparent Extract– Final Gravity (°Plato):	1.014–1.018 (3.5–4.5 °Plato)
Alcohol by weight (volume):	3.8–4.2% (4.5–5%)
Bitterness (IBU):	16–25
Color (SRM):	17–20 (40–80 EBC)

B. SUBCATEGORY: GERMAN-STYLE SCHWARZBIER

These beers have a roasted-malt character without the associated bitterness. Malt flavor and aroma are low in sweetness. Hop bitterness is low to medium in character. Noble-type hop flavor and aroma should be low but perceptible. There should be no fruity esters. Diacetyl is acceptable at very low levels.

Original Gravity (°Plato):	1.044–1.052 (11–13 °Plato)
Apparent Extract– Final Gravity (°Plato):	1.012–1.016 (3–4 °Plato)
Alcohol by weight (volume):	3–3.9% (3.8–5%)
Bitterness (IBU):	22–30
Color (SRM):	25–30 (100–120 EBC)

Gold Cup Winner

Tabernash Munich Dark Lager
Tabernash Brewing Company
205 Denargo Market
Denver, Colorado, USA 80216
Brewmaster: Eric Warner
Established 1993
Production: 8,000 bbl. (9,400 hl.)

Colorado is the land of mountains and BEER. At this writing there are close to 100 breweries in a state inhabited by fewer than 3 million people. That's one brewery for every 30,000 people. Though breweries proliferate in Colorado, Tabernash Brewing Company was alone when it established itself as a lager microbrewery, specializing in traditional German-style lagers and Bavarian-style wheat beers. Eric Warner, one of the founders, is a brewmaster trained in the

Eric Warner

Eric Warner, Brewmaster

tradition of the famed Technical University of Munich at Weihenstephan. Eric's brewing philosophy inspires the importation of malt and hops from Germany and special strains of yeast from the world's oldest brewery, Bavaria's Weihenstephan, which has been brewing since 1040. Eric Warner is also the author of *German Wheat Beer* (Brewers Publications, 1992). This introduction wouldn't be complete without a mention that Eric Warner, too, began as a homebrewer.

Tabernash beers are not strangers to fame and have claimed numerous medals over the years at the Great American Beer Festival in professional panel blind tastings. The brewery regularly brews four beers, Tabernash Weiss, Tabernash Golden, Tabernash Amber and Tabernash Munich, as well as seasonal beers such as Doppelbock, Oktoberfest and a winter specialty.

CHARACTER DESCRIPTION OF GOLD CUP–WINNING TABERNASH
MUNICH DARK LAGER

Chestnut brown lager with a transparent ruby red hue. Malty aroma has some toasted-malt character. A hint of fruitiness; almost applelike. Hop aroma is mostly lacking, emerging only enough to be evident. Malt aroma predominates, but all aromatics are subdued, seeming to invite, "Seek me . . . discover more of me . . . explore me. . . . There's more of me, but I'm not easy to get to. . . ." Light to medium body is unexpected given the dark color and malty aroma. Overall sweet malt character in flavor does not yield to hop bitterness, which is so low there is virtually no bitterness in aftertaste. A hint of fruitiness complements an overall balance of chocolate, banana and malt, accented by the low level of bitterness. Roast malt contributes to a hint of acidity, not to be confused with bitterness. Very low but perceptible level of diacetyl rounds out the character. Aftertaste is clean and neutral.

- **Recipe for 5 U.S. gallons (19 liters) Tabernash Munich Dark Lager**
  Targets:
  Original Gravity: 1.048 (12)
  Final Gravity: 1.014 (3.5)
  Alcohol by volume: 4.5%
  Color: 18 SRM (36 EBC)
  Bittering Units: 17

ALL-GRAIN RECIPE AND PROCEDURE

3 ¾ lbs.	(1.7 kg.)	German Munich malt—12 Lovibond (20–25 EBC)
4 ¾ lbs.	(2.2 kg.)	Canadian Munich malt—15 Lovibond (30 EBC)
¼ lb.	(114 g.)	German CaraMunich—20 Lovibond (40 EBC)
¼ lb.	(114 g.)	American Victory malt, Biscuit malt or other aromatic malt
2 ½ oz.	(70 g.)	chocolate malt
9.2 lbs.	**(4.2 kg.)**	**Total grains**

2	HBU	(56 MBU) German Northern Brewer hops (pellets)—75 minutes (bittering)
3	HBU	(85 MBU) German Perle hops (pellets)—30 minutes (flavor)
1	HBU	(28 MBU) German Hersbrucker Hallertauer hops (pellets)—15 minutes (flavor)

¼	tsp.	Irish moss
¾	c.	corn sugar for priming in bottles. Use ⅓ cup corn sugar if priming a keg.

Wyeast 2278 Czech Pils yeast or Weihenstephan 34–70 Lager yeast; medium attenuation, clean, dry but malty finish.

A step infusion mash is employed to mash the grains. Add 9 quarts (8.5 l.) of 140-degree F (60 C) water to the crushed grain, stir, stabilize, and hold the temperature at 133 degrees F (56 C) for 30 minutes. Add 4.5 quarts (4.3 l.) of boiling water, adding heat if necessary to bring temperature up to 155 degrees F (68 C). Hold for about 60 minutes.

After conversion, raise temperature to 167 degrees F (75 C), lauter and sparge with 4 gallons (15 l.) of 170-degree F (77 C) water. Collect about 6 gallons (23 l.) of runoff, add bittering hops and bring to a full and vigorous boil.

The total boil time will be 75 minutes. When 30 minutes remain, add flavor hops. When 15 minutes remain, add aroma hops and Irish moss. After a total wort boil of 75 minutes (reducing the wort volume to just over 5 gallons), turn off the heat, then separate or strain out and sparge hops. Chill the wort to 65 degrees F (18 C) and direct into a sanitized fermenter. Aerate the cooled wort well. Add an active yeast culture and ferment for 4 to 6 days in the primary at 55 degrees F (15 C). Then transfer into a secondary fermenter, maintain 55 degrees F (15 C) to age for two more weeks, then lager for two to four more weeks at 40 degrees F (4.5 C).

When secondary aging is complete, prime with sugar, bottle or keg. Let condition at temperatures above 60 degrees F (15.5 C) until clear and carbonated, then store chilled.

MASH-EXTRACT RECIPE AND PROCEDURE FOR TABERNASH MUNICH DARK LAGER

3 ¾ lbs.	(1.7 kg.)	English light dried malt extract
1 ¼ lbs.	(0.6 kg.)	German Munich malt—12 Lovibond (20–25 EBC)
1 ¼ lbs.	(0.6 kg.)	Canadian Munich malt—15 Lovibond (30 EBC)
¼ lb.	(114 g.)	American Victory malt, Biscuit malt or other aromatic malt
¼ lb.	(114 g.)	chocolate malt
3 lbs.	**(1.4 kg.)**	**Total grains**

3	HBU	(85 MBU) German Northern Brewer hops (pellets)—75 minutes (bittering)
3.5	HBU	(99 MBU) German Perle hops (pellets)—30 minutes (flavor)
1	HBU	(28 MBU) German Hersbrucker Hallertauer hops (pellets)—15 minutes (flavor)

¼	tsp.	Irish moss
¾	c.	corn sugar for priming in bottles. Use ⅓ cup corn sugar if priming a keg.

Wyeast 2278 Czech Pils yeast or Weihenstephan 34–70 Lager yeast; medium attenuation, clean, dry but malty finish.

A step infusion mash is employed to mash the grains. Add 3 quarts (2.8 l.) of 140-degree F (60 C) water to the crushed grain, stir, stabilize and hold the temperature at 133 degrees F (56 C) for 30 minutes. Add 1.5 quarts (1.5 l.) of boiling water and add heat to bring temperature up to 155 degrees F (68 C). Hold for about 60 minutes.

After conversion, raise temperature to 167 degrees F (75 C), lauter and sparge with 2 gallons (7.6 l.) of 170-degree F (77 C) water. Collect about 2.5 gallons (9.5 l.) of runoff. Add malt extract and bittering hops and bring to a full and vigorous boil.

The total boil time will be 75 minutes. When 30 minutes remain, add flavor hops. When 15 minutes remain, add aroma hops and Irish moss. After a total wort boil of 75 minutes, turn off the heat, then separate or strain out and sparge hops, and direct the hot wort into a sanitized fermenter to which 2 gallons (7.6 l.) of cold water have been added. If necessary, add additional cold water to achieve a 5-gallon (19-l.) batch size. Chill the wort to 65 degrees F (68.5 C). Aerate the cooled wort well. Add an active yeast culture and ferment for 4 to 6 days in the primary at 55 degrees F (15 C). Then transfer into a secondary fermenter, maintain 55 degrees F (15 C) to age for two more weeks, then lager for two to four more weeks at 40 degrees F (4.5 C).

When secondary aging is complete, prime with sugar, bottle or keg. Let condition at temperatures above 60 degrees F (15.5 C) until clear and carbonated, then store chilled.

Silver Cup Winner

Blue Hen Black & Tan
Blue Hen Beer Co. Ltd.
Newark, Delaware, USA

Deep brown color with ruby red hue. Roasted malt is primary aroma, with notes of cocoa and toasted malt. Dark caramel may also be involved. Medium-bodied mouth feel with an elegant full malt flavor. Clean with great balance between malt flavor and an evident hop bitterness that yields only briefly to roast-malt bitterness. A relatively dry finish as maltiness quickly fades to a pleasant cocoa and hop bitterness. Bitterness offers a harsh impact rather than the soft bitterness usually expected in this style. A blend of "lager and porter" as stated on the label, and passing for a European dark lager due to the cleanness of the beer, the texture of the roasted malts, the dry finish and adequate bitterness. Medium-bodied but tends toward a dry and light finish. This beer is more indicative of the hoppier and drier North German styles of dark lagers.

Estimated profile based on tasting
Color 22–25 SRM (44–50 EBC)
Bittering Units: 34–36

Bronze Cup Winner

Stovepipe Porter
Otter Creek Brewing Inc.
Middlebury, Vermont, USA

Dark brown ale with a deep ruby red hue. Cocoa, caramel cookie and biscuit aromas predominate. No notable hop aroma. A subdued but evident fruity complex ale character reveals the top fermentation of this ale passed as a lager. But overall character is neutral and dominated by malt rather than fermentation character. Medium-bodied mouth feel while malt flavors mirror aroma. Hop bitterness is balanced and not overstated, complemented by the zesty astringency of roasted malts. A great combination, trailing off with a pleasing dryness in the aftertaste. Notably clean with great balance between malt sweetness and hop bitterness.

Estimated profile based on tasting
Color: 18–21 SRM (36–42 EBC)
Bittering Units: 30–32

CATEGORY 44:
GERMAN-STYLE BOCK BEER

A. SUBCATEGORY: TRADITIONAL BOCK

Traditional Bocks are made with all malt and are strong, malty, medium- to full-bodied, bottom-fermented beers with moderate hop bitterness that should increase proportionately with the starting gravity. Hop flavor should be low, and hop aroma should be very low. Bocks can range in color from deep copper to dark brown. Fruity esters may be perceived at low levels.

Original Gravity (°Plato):	1.066–1.074 (16.5–18.5 °Plato)
Apparent Extract– Final Gravity (°Plato):	1.018–1.024 (4.5–6 °Plato)
Alcohol by weight (volume):	5–6% (6–7.5%)
Bitterness (IBU):	20–30
Color SRM (EBC):	20–30 (70–120 EBC)

B. SUBCATEGORY: GERMAN-STYLE HELLES BOCK/MAIBOCK

The German word *hell* means light-colored, and accordingly, a Helles Bock is light in color. Maibocks are light-colored bocks. The malty character should come through in the aroma and flavor. Body is medium to full. Hop bitterness should be low while noble hop aroma and flavor may be at low to medium levels. Bitterness increases with gravity. Fruity esters should be minimal. Diacetyl levels should be very low. Chill haze should not be perceived.

Original Gravity (°Plato):	1.066–1.068 (16.5–17 °Plato)
Apparent Extract–Final Gravity (°Plato):	1.012–1.020 (3–5 °Plato)
Alcohol by weight (volume):	5–6% (6–7.5%)
Bitterness (IBU):	20–35
Color SRM (EBC):	4–10 (10–20 EBC)

Gold Cup Winner

Ruffian Mai-Bock
Mountain Valley Brew Pub
122 Orange Ave.
Suffern, New York, USA 10901
Brewmaster: Jay Misson
Established 1992
Production: 4,000 bbl. (4,700 hl.)

Mountain Valley also won the Bohemian-Style Pilsener Category. Please see that Category for a description of the brewery.

CHARACTER DESCRIPTION OF GOLD CUP–WINNING RUFFIAN MAI-BOCK

Mountain Valley Brew Pub, Home of:

Ruffian
Ales and Lagers ™

Lon M. Lauterio

Dense, rich, thick, white head atop a light golden amber vein of gold. Aroma evolves to the rich, clean aroma of pure beer with subtle malt and a very subtle berrylike fruity ester. Hop aroma is absent. Overall a very neutral malt-accented aroma. Flavor is a powerful wave of unsatiating pale malt sweetness. Munich, caramel and chocolate malt characters are totally absent. Bitterness is balanced and soft with a refreshing bite in the aftertaste to cleanse the sweetness. Medium-bodied mouth feel; it must be the throat-warming alcohol that causes a paradoxically full-bodied swallow. The cleanness of character and purity of alcohol indicate a controlled and well-choreographed fermentation. Ample lagering has provided award-winning smoothness. A beer any German (or American) would be proud to brew. A beer of dreams, slumber and coveted memory.

- ### Recipe for 5 U.S. gallons (19 liters) Ruffian Mai-Bock

  Targets:
  Original Gravity: 1.065 (16)
  Final Gravity: 1.014 (3.5)
  Alcohol by volume: 6.7%
  Color: 7 SRM (14 EBC)
  Bittering Units: 27

ALL-GRAIN RECIPE AND PROCEDURE

11½ lbs.	(5.2 kg.)	German Pilsener 2-row malt
¼ lb.	(114 g.)	German Sauer (sour) malt
¼ lb.	(114 g.)	English crystal malt—10 Lovibond
12 lbs.	**(5.4 kg.)**	**Total grains**

3	HBU	(85 MBU) German Spalt hops (pellets)—105 minutes (bittering)
2	HBU	(56 MBU) German Tettnanger hops (pellets)—105 minutes (bittering)
2	HBU	(56 MBU) German Tettnanger hops (pellets)—30 minutes (flavor)
1	HBU	(28 MBU) Czech Saaz hops (pellets)—10 minutes (aroma)

¼   tsp.   Irish moss
¾   c.   corn sugar for priming in bottles. Use ⅓ cup
            corn sugar if priming a keg.
Wyeast 2206 Bavarian Lager yeast or Wyeast 2308 Munich
Lager yeast

A double-decoction mash is employed to mash the grains.
If you are using alkaline or hard water, treat the water with
an appropriate amount of lactic acid or refer to pages 120
to 122 of *The Home Brewer's Companion* for proceeding with a
triple decoction with an acid rest.

Add 12 quarts (11.4 l.) of boiling water to the grain, raising the temperature to 122 degrees F (50 C). Then remove
6 quarts (5.7 l.) of the thickest mash and boil this in another
vessel for 30 minutes. Stir this boiled decoction constantly.
Maintain the remainder of the mash at 122 degrees F (50
C) while boiling the decocted portion. After boiling for 30
minutes, add the decoction back into the main mash vessel,
raising the temperature to 150 degrees F (65.5 C). Maintain
this temperature for one hour. Up to 12 quarts (8.6 l.) of
boiling water may be added at any time during this period
to maintain mashing temperature. After one hour, remove
one third of the mash (a fifty-fifty blend of thick mash and
liquid) and boil about 20 minutes. Stir this second decoction constantly during the boil. When finished, return the
boiled decoction to the main mash vessel, ending starch
conversion by having raised the temperature to about 167
degrees F (75 C). Then lauter and sparge with 5 gallons (19
l.) of 170-degree F (77 C) water. Collect about 7 gallons (27
l.) of runoff, add bittering hops and bring to a full and
vigorous boil.

The total boil time will be 105 minutes. When 30 minutes
remain, add flavor hops. When 10 minutes remain, add
aroma hops and Irish moss. After a total wort boil of 105
minutes (reducing the wort volume to just over 5 gallons),
turn off the heat, then separate or strain out and sparge
hops. Chill the wort to 65 degrees F (18 C) and direct into

a sanitized fermenter. Aerate the cooled wort well. Add an active yeast culture and ferment for 4 to 6 days in the primary at 55 degrees F (15 C). Then transfer into a secondary fermenter, chill to 50 degrees F (10 C) to age for two more weeks, then lager for four or more weeks at 40 degrees F (4.5 C).

When secondary aging is complete, prime with sugar, bottle or keg. Let condition at temperatures above 60 degrees F (15.5 C) until clear and carbonated, then store chilled.

MASH-EXTRACT RECIPE AND PROCEDURE FOR RUFFIAN MAI-BOCK

5	lbs.	(2.3 kg.)	English light dried malt extract
3 ½	lbs.	(1.6 kg.)	German Pilsener 2-row malt
¼	lb.	(114 g.)	German Sauer (sour) malt
¼	lb.	(114 g.)	English crystal malt—10 Lovibond
4	**lbs.**	**(1.8 kg.)**	**Total grains**

4	HBU	(113 MBU) German Spalt hops (pellets)—60 minutes (bittering)
3	HBU	(85 MBU) German Tettnanger hops (pellets)—60 minutes (bittering)
2	HBU	(56 MBU) German Tettnanger hops (pellets)—30 minutes (flavor)
1	HBU	(28 MBU) Czech Saaz hops (pellets)—10 minutes (aroma)

¼	tsp.	Irish moss
¾	c.	corn sugar for priming in bottles. Use ⅓ cup corn sugar if priming a keg.

Wyeast 2206 Bavarian Lager yeast or Wyeast 2308 Munich Lager yeast.

A double-decoction mash is employed to mash the grains. If you are using alkaline or hard water, treat the water with an appropriate amount of lactic acid or refer to pages 120 to 122 of *The Home Brewer's Companion* for proceeding with a triple decoction with an acid rest.

Add 4 quarts (3.8 l.) of boiling water to the grain, raising the temperature to 122 degrees F (50 C). Then remove 2 quarts (1.9 l.) of the thickest mash and boil this in another vessel for 30 minutes. Stir this boiled decoction constantly. Maintain the remainder of the mash at 122 degrees F (50 C) while boiling the decocted portion. After boiling for 30 minutes, add the decoction back into the main mash vessel, raising the temperature to 150 degrees F (65.5 C). Maintain this temperature for one hour. Up to 4 quarts (8.6 l.) of boiling water may be added at any time during this period to maintain mashing temperature. After one hour, remove one third of the mash (a fifty-fifty blend of thick mash and liquid) and boil about 20 minutes. Stir this second decoction constantly during the boil. When finished, return the boiled decoction to the main mash vessel, ending starch conversion by having raised the temperature to about 167 degrees F (75 C). Then lauter and sparge with 2 gallons (7.6 l.) of 170-degree F (77 C) water. Collect about 3 gallons (11.4 l.) of runoff, add bittering hops and bring to a full and vigorous boil.

The total boil time will be 60 minutes. When 30 minutes remain, add flavor hops. When 10 minutes remain, add aroma hops and Irish moss. After a total wort boil of 60 minutes, turn off the heat, separate or strain out and sparge hops, and direct the hot wort into a sanitized fermenter to which 2 gallons (7.6 l.) of cold water have been added. If necessary, add additional cold water to achieve a 5-gallon (19-l.) batch size. Chill the wort to 65 degrees F (18 C) and direct into a sanitized fermenter. Aerate the cooled wort well. Add an active yeast culture and ferment for 4 to 6 days in the primary at 55 degrees F (15 C). Then transfer into a secondary fermenter, chill to 50 degrees F (10 C) to age for two more weeks, then lager for four or more weeks at 40 degrees F (4.5 C).

When secondary aging is complete, prime with sugar, bottle or keg. Let condition at temperatures above 60 degrees F (15.5 C) until clear and carbonated, then store chilled.

Silver Cup Winner

Hübsch Mai Bock
Sudwerk Privatbrauerei Hübsch
Davis, California, USA

Amber color with an orange hue. Aroma immediately suggests an exotic and unusual woody, cedarlike character. Flavor follows with a full impact of malt supported by a medium-bodied mouth feel. Pleasant, soft hop bitterness follows and balances malt sweetness. Slight fruitiness, but malt sweetness and hop bitterness interplay and dominate. Caramel-type sweetness is absent. Amber color is likely not from caramel or aromatic malt but from Munich or Vienna-style malt, also lending a chewiness to flavor and aftertaste. Another very clean beer with malt and hops as the overwhelming thematic impression. Fruitiness from such a strong beer is barely evident. Smooth and well-lagered character is evident. Though substantially bittered for balance, the overall impression is a sweet, chewy, mild yet austere malt character.

Estimated profile based on tasting
Color: 10–12 SRM (20–24 EBC)
Bittering Units: 28–31

Bronze Cup Winner

Augsburger Dopplebock
Augsburger Brewing Co.
Detroit, Michigan, USA

Rich, dark brown color with glimmer of ruby red. Aroma portrays shades of roasted malt and cocoa without being overstated. The malt aroma is rich. It does not reveal a strong caramel-malt character, though this malt is probably used. High carbonation is memorable and influences overall flavor characters by suggesting a well-attenuated dry fermentation. Still, the coca reveals itself, coming through with a rich blend of specialty malts. Medium-bodied and by all flavor indications not quite up to the strength of a traditional German double bock. It's all on the right frequency, but the volume is not turned up quite high enough. Hop bitterness is in balance with malt and is not overstated; perfect for style. A bit too carbonated and dry in the finish, making it quite drinkable and refreshing, but a double bock is not intended to refresh the thirst, it is meant to refresh the mind. The tingle of alcohol is absent.

Brewery formulation uses caramel and chocolate malts with American Willamette and Mt. Hood hops for bitterness, flavor and aroma.

Estimated profile based on tasting
Original Gravity: 1.056 (14) indicated by the brewery
Final Gravity: 1.027 (6.7) indicated by the brewery
Alcohol by volume: 5% indicated by the brewery
Color: 18–20 SRM (36–40 EBC) (70 SRM indicated by the brewery)
Bittering Units: 21–25 (20 indicated by the brewery)

CATEGORY 45:
GERMAN-STYLE STRONG BOCK BEER

Malty sweetness is dominant, but should not be cloying. Doppelbocks are full-bodied and deep amber to dark brown in color. Astringency from roast malts is absent. Alcoholic strength is high, and hop rates increase with gravity. Hop bitterness and flavor should be low, with hop aroma absent. Fruity esters are commonly perceived, but at low to moderate levels.

Original Gravity (°Plato):	1.074–1.080 (18.5–20 °Plato)
Apparent Extract–	
Final Gravity (°Plato):	1.020–1.028 (5–7 °Plato)
Alcohol by weight (volume):	5.2–6.2% (6.5–8%)
Bitterness (IBU):	17–27
Color SRM (EBC):	12–30 (30–120 EBC)

Gold Cup Winner

Tabernash Doppelbock
Tabernash Brewing Company
205 Denargo Market
Denver, Colorado, USA 80216
Brewmaster: Eric Warner
Established 1993
Production: 8,000 bbl. (9,400 hl.)

Tabernash also won the European-Style Dark Lager category.
Please see that Category for a description of the brewery.

CHARACTER DESCRIPTION OF GOLD CUP–WINNING TABERNASH
DOPPELBOCK

A medium dark brown lager with a red hue. Aroma blossoms with a soft, delicate ester-and-malt aroma reminiscent of cocoa and hints of caramel toffee. Flavor presents itself as memorably sweet with malt and big coffee-toffee notes. Body is full to medium, subduing the assaulting sensation of alcohol, though alcohol warms on its postpalate journey. Hop bitterness is soft, coming on with help from roast-malt dryness, both quickly yielding to the overall sensation of gentle sweetness. Aftertaste of dry cocoalike character revisits with a subdued bitterness and palate-cleansing sensation.

Eric Warner

- **Recipe for 5 U.S. gallons (19 liters) Tabernash Doppelbock**

Targets:
Original Gravity: 1.074 (18)
Final Gravity: 1.026 (6.5)
Alcohol by volume: 6.5%
Color: 25 SRM (50 EBC)
Bittering Units: 29

ALL-GRAIN RECIPE AND PROCEDURE

6¼ lbs.	(2.8 kg.)	German Munich malt—12 Lovibond (20–25 EBC)
6¾ lbs.	(3.1 kg.)	Canadian Munich malt—15 Lovibond (30 EBC)
½ lb.	(0.23 kg.)	German CaraMunich malt—20 Lovibond (40 EBC)
¼ lb.	(114 g.)	American Victory malt, Biscuit malt or other aromatic malt
1½ oz.	(42 g.)	chocolate malt
13¾ lbs.	**(6.2 kg.)**	**Total grains**

6.5 HBU (184 MBU) German Northern Brewer hops (pellets)—105 minutes (bittering)

¾ oz. (21 g.) German Hersbrucker Hallertauer hops (pellets)—steep in finished boiled wort for 2 to 3 minutes (aroma)

¼ tsp. Irish moss
¾ c. corn sugar for priming in bottles. Use ⅓ cup corn sugar if priming a keg.

Wyeast 2278 Czech Pils yeast or Weihenstephan 34–70 Lager yeast; medium attenuation, clean, dry but malty finish.

A step infusion mash is employed to mash the grains. Add 14 quarts (13 l.) of 140-degree F (60 C) water to the crushed grain, stir, stabilize and hold the temperature at 133 degrees F (56 C) for 30 minutes. Add 7 quarts (6.7 l.) of boiling water, adding heat if necessary to bring temperature up to 150 degrees F (65.5 C). Hold for about 60 minutes.

After conversion, raise temperature to 167 degrees F (75 C), lauter and sparge with 4 gallons (15 l.) of 170-degree F (77 C) water. Collect about 7.5 gallons (28.5 l.) of runoff, add bittering hops and bring to a full and vigorous boil.

The total boil time will be 105 minutes. When 10 minutes remain, add Irish moss. After a total wort boil of 105 minutes (reducing the wort volume to just over 5 gallons), turn off the heat, add aroma hops and let steep for 2 to 3 minutes, then separate or strain out and sparge hops. Chill the wort to 65 degrees F (18 C) and direct into a sanitized fermenter. Aerate the cooled wort well. Add an active yeast culture and ferment for 4 to 6 days in the primary at 55 degrees F (15 C). Then transfer into a secondary fermenter, chill to 50 degrees F (10 C) to age for two more weeks, then lager for four to six more weeks at 40 degrees F (4.5 C).

When secondary aging is complete, prime with sugar, bottle or keg. Let condition at temperatures above 60 degrees F (15.5 C) until clear and carbonated, then store chilled.

MASH-EXTRACT RECIPE AND PROCEDURE FOR TABERNASH DOPPELBOCK

5	lbs.	(2.3 kg.)	English amber dried malt extract
2½	lbs.	(1.1 kg.)	German Munich malt—12 Lovibond (20–25 EBC)
2½	lbs.	(1.1 kg.)	Canadian Munich malt—15 Lovibond (30 EBC)
½	lb.	(0.23 kg.)	German CaraMunich malt—20 Lovibond (40 EBC)
¼	lb.	(114 g.)	American Victory malt, Biscuit malt or other aromatic malt
2½	oz.	(70 g.)	chocolate malt
6	**lbs.**	**(2.7 kg.)**	**Total grains**

8.5   HBU   (241 MBU) German Northern Brewer hops
            (pellets)—75 minutes (bittering)
¾   oz.   (21 g.) German Hersbrucker Hallertauer hops
            (pellets)—steep in finished boiled wort for 2
            to 3 minutes (aroma)

¼   tsp.   Irish moss
¾   c.   corn sugar for priming in bottles. Use ⅓ cup
            corn sugar if priming a keg.
Wyeast 2278 Czech Pils yeast or Weihenstephan 34–70
Lager yeast; medium attenuation, clean, dry but malty
finish.

A step infusion mash is employed to mash the grains.
Add 6 quarts (5.7 l.) of 140-degree F (60 C) water to the
crushed grain, stir, stabilize and hold the temperature at
133 degrees F (56 C) for 30 minutes. Add 3 quarts (2.9 l.)
of boiling water. Add heat to bring temperature up to 150
degrees F (65.5 C). Hold for about 60 minutes.

After conversion, raise temperature to 167 degrees F (75
C), lauter and sparge with 2 gallons (7.6 l.) of 170-degree F
(77 C) water. Collect about 3 gallons (11.5 l.) of runoff. Add
malt extract and bittering hops and bring to a full and vigorous boil.

The total boil time will be 75 minutes. When 10 minutes
remain, add Irish moss. After a total wort boil of 105 minutes (reducing the wort volume to just over 5 gallons), turn
off the heat, add aroma hops and let steep for 2 to 3 minutes. Then separate or strain out and sparge hops, and direct the hot wort into a sanitized fermenter to which 2
gallons (7.6 l.) of cold water have been added. If necessary,
add additional cold water to achieve a 5-gallon (19-l.) batch
size. Chill the wort to 70 degrees F (21 C). Aerate the cooled
wort well. Add an active yeast culture and ferment for 4 to
6 days in the primary at 55 degrees F (15 C). Then transfer
into a secondary fermenter, chill to 50 degrees F (10 C) to

age for two more weeks, then lager for four to six more weeks at 40 degrees F (4.5 C).

When secondary aging is complete, prime with sugar, bottle or keg. Let condition at temperatures above 60 degrees F (15.5 C) until clear and carbonated, then store chilled.

Silver Cup Winner

Scapegoat Doppelbock
Libertyville Brewing Co.
Libertyville, Illinois, USA

Rich brown color with hints of red and orange. It's all malt in the aroma; rich, full and exceptionally authentic. Forget the hop aroma, as one shouldn't expect it with this style. Am I in Germany? Alcohol aroma is absent, but a hint of high-gravity, lagered fruitiness is welcoming in aroma and flavor. Full-bodied mouth feel with rich malt character. Roasted malts are not at all astringent, nor do they add bitterness. Mellow, well-lagered flavor. Hops are present for balance. Low bitterness totally yields to the rich malt character, though without the supporting role of hops, the rich malt character would be overwhelmingly satiating. A devilishly quiet, clean aftertaste, smooth, velvety and soothing. The skillful use of malts and lagering says it all.

Brewery formulation uses Munich, Crystal, dextrine and other specialty malts, with German Perle and American Liberty hops for bitterness and flavor.

Estimated profile based on tasting
Original Gravity: 1.076 (18.6) indicated by the brewery
Final Gravity: 1.021 (5.2) indicated by the brewery
Alcohol by volume: 8.5% indicated by the brewery
Color: 21–24 SRM (42–48 EBC) (30 indicated by brewery)
Bittering Units: 24–30 (17 indicated by the brewery)

Bronze Cup Winner

Bayrisch G. Frorns Eisbock
Kulmbacher Reichelbraeu
Kulmbach, Germany

A big, bold pour reveals bubbles struggling to reach the surface. Deep chocolate brown color with definitive ruby red hues. Aroma reveals full though not overpowering malt aroma. Gentle softness of roasted malts suggest a subtle mix of cocoa and toasted-malt characters. Full bready malt aroma so very characteristic of German-style decoction mashes. The aroma is a very important part of the wonderful experience this beer offers. Full-bodied mouth feel and full malt flavor. Subtle bitterness of hops is certainly evident, but only is there to balance malt richness. A very slight DMS (sweet cornlike character) in flavor helps accent the malt. High alcohol content is devilishly deceiving, as the full malt character masks the warmth created by the alcohol. No harshness to be experienced anywhere. A gentle beer with a tendency to induce dreamlike states of mind.

Brewery formulation uses German dark, Vienna, caramel and crystal malts, with Brewers Gold, Perle, Hersbrucker and Tettnanger hops.

Estimated profile based on tasting
Original Gravity: 1.100 (24) indicated by the brewery
Final Gravity: 1.026 (6.5)
Alcohol by volume: 10% indicated by the brewery
Color: 25–29 SRM (50–58 EBC) (130 EBC indicated by the brewery)
Bittering Units: 28–32 (28 indicated by the brewery)

North American Origin

CATEGORY 46:
AMERICAN-STYLE LAGER

Very light in body and color, American Lagers are very clean and crisp and aggressively carbonated. Malt sweetness is absent. Corn, rice and other grain or sugar adjuncts are often used. Hop aroma is absent. Hop bitterness is slight and hop flavor is mild or negligible. Chill haze, fruity esters and diacetyl should be absent.

Original Gravity (°Plato):   1.040–1.046 (10–11.5 °Plato)
Apparent Extract–
Final Gravity (°Plato):   1.006–1.010 (1.5–2.5 °Plato)
Alcohol by weight (volume): 3.2–3.8% (3.8–4.5%)
Bitterness (IBU):   5–17
Color (SRM):   2–4 (4–7 EBC)

Gold Cup Winner

OB Lager
Oriental Brewery Co.
Technical Center
Sungbokri 39-3,
Suji Eup, Yongin
Kyunggi Province, Korea
Head Brewer: In Woo Yoon, executive managing director,
Technical Center
Brewmaster: Young Kyu Kim, director, Kumi Brewery
Established 1952
Production: 7.14 million bbl. (8.36 million hl.)

The largest brewing company in Korea, the Oriental Brewery Co. now has four brewing facilities and two malting plants. A fifth brewery is under construction, attesting to the beer's popularity both in and outside of Korea; the brewery exports to Japan, Hong Kong, Europe and the United States. Its main products are light lagers and include OB Lager, Nex, Cafri, OB Light, OB Draft and OB Sound.

Mr. In Woo Yoon, Head Brewer   *Mr. Young Kyu Kim, Brewmaster*

Jung Hyun Ahn

CHARACTER DESCRIPTION OF GOLD CUP–WINNING OB LAGER

Extremely pale and crystal-clear. Very subtle sulfur character expresses itself alongside an otherwise very clean and neutral aroma. Barely evident is the pleasant herbal-minty aroma of noble hops. Noble hop character is also evident in flavor, which is otherwise exceptionally clean and crisp, complemented with a light body. Hop bitterness is in balance and seems to express itself a bit more than it does in popular American light lagers, but analysis shows this is just a pleasant illusion. Smooth and well-lagered character. A beer that goes down very easily, with the added attraction of being balanced with a perfect addition of hop bitterness for the style. No acidic aftertaste as with many underhopped American-style lagers. Admirably clean aftertaste.

• **Recipe for 5 U.S. gallons (19 liters) OB Lager**
Targets:
Original Gravity: 1.042 (10.5)
Final Gravity: 1.008 (2)
Alcohol by volume: 4.6%
Color: 2.5 SRM (5 EBC)
Bittering Units: 13

ALL-GRAIN RECIPE AND PROCEDURE

2¼ lbs.	(1 kg.)	Canadian 2-row Harrington pale malt
3¼ lbs.	(1.5 kg.)	Korean Doosan 2-row (or American Klages) pale malt
2½ lbs.	(1.1 kg.)	flaked corn
8 lbs.	**(3.6 kg.)**	**Total grains**

1.5	HBU	(43 MBU) American Cluster hops (pellets)—75 minutes (bittering)
1	HBU	(28 MBU) American Nugget hops (pellets)—75 minutes (bittering)

½   oz.   (14 g.) German Hersbrucker Hallertauer hops
(pellets)—steep in finished boiled wort for 2
to 3 minutes (aroma)

¼   tsp.   Irish moss
¾   c.   corn sugar for priming in bottles. Use ⅓ cup
corn sugar if priming a keg.
Wyeast 2007 Pilsen Lager yeast

A step infusion mash is employed to mash the grains.
Add 8 quarts (7.6 l.) of 130-degree F (54.5 C) water to the
crushed grain and flaked corn, stir, stabilize and hold the
temperature at 122 degrees F (50 C) for 30 minutes. Add 4
quarts (3.8 l.) of boiling water, adding heat if necessary to
bring temperature up to 150 degrees F (65.5 C). Hold for
about 60 minutes.

After conversion, raise temperature to 167 degrees F
(75 C), lauter and sparge with 4 gallons (15 l.) of 170-
degree F (77 C) water. Collect about 6 gallons (23 l.) of
runoff, add bittering hops and bring to a full and vigorous boil.

The total boil time will be 75 minutes. When 10 minutes
remain, add Irish moss. After a total wort boil of 75 minutes
(reducing the wort volume to just over 5 gallons), turn off
the heat, add the aroma hops and let steep for 3 to 5 minutes, then separate or strain out and sparge hops. Chill the
wort to 65 degrees F (18 C) and direct into a sanitized fermenter. Aerate the cooled wort well. Add an active yeast
culture and ferment for 4 to 6 days in the primary at 50
degrees F (10 C). Then transfer into a secondary fermenter.
Maintain temperature of 50 degrees F (10 C) to age for two
more weeks, then lager for two to four more weeks at 40
degrees F (4.5 C).

When secondary aging is complete, prime with sugar, bottle
or keg. Let condition at temperatures above 60 degrees F
(15.5 C) until clear and carbonated, then store chilled.

MASH-EXTRACT RECIPE AND PROCEDURE FOR OB LAGER

2¼ lbs. (1 kg.)   English light dried malt extract

2½ lbs. (1.1 kg.)   American Harrington/Klages 2-row pale malt

1½ lbs. (0.7 kg.)   flaked corn

4   lbs. (1.8 kg.)   Total grains

2   HBU   (56 MBU) American Cluster hops (pellets)—75 minutes (bittering)

1   HBU   (28 MBU) American Nugget hops (pellets)—75 minutes (bittering)

½   oz.   (14 g.) German Hersbrucker Hallertauer hops (pellets)—steep in finished boiled wort for 2 to 3 minutes (aroma)

¼   tsp.   Irish moss

¾   c.   corn sugar for priming in bottles. Use ⅓ cup corn sugar if priming a keg.

Wyeast 2007 Pilsen Lager yeast

A step infusion mash is employed to mash the grains. Add 4 quarts (3.8 l.) of 130-degree F (54.5 C) water to the crushed grain and flaked corn, stir, stabilize and hold the temperature at 122 degrees F (50 C) for 30 minutes. Add 2 quarts (1.9 l.) of boiling water. Add heat to bring temperature up to 150 degrees F (65.5 C). Hold for about 60 minutes.

After conversion, raise temperature to 167 degrees F (75 C), lauter and sparge with 2 gallons (7.6 l.) of 170-degree F (77 C) water. Collect about 3 gallons (11.5 l.) of runoff. Add malt extract and bittering hops and bring to a full and vigorous boil.

The total boil time will be 75 minutes. When 10 minutes remain, add Irish moss. After a total wort boil of 75 minutes, turn off the heat, add aroma hops and let steep 2 to 5 minutes. Then separate or strain out and sparge hops, and direct the hot wort into a sanitized fermenter to which 2 gallons (7.6 l.) of cold water have been added. If necessary, add additional cold water to achieve a 5-gallon (19-l.) batch size. Chill the wort to 65 degrees F (18 C) and direct into a sanitized fermenter. Aerate the cooled wort well. Add an active yeast culture and ferment for 4 to 6 days in the primary at 50 degrees F (10 C). Then transfer into a secondary fermenter. Maintain temperature of 50 degrees F (10 C) to age for two more weeks, then lager for two to four more weeks at 40 degrees F (4.5 C).

When secondary aging is complete, prime with sugar, bottle or keg. Let condition at temperatures above 60 degrees F (15.5 C) until clear and carbonated, then store chilled.

Silver Cup Winner

Nex
Oriental Brewery Co. Ltd.
Seoul, Korea

Pale straw in color. Great fresh beer aroma with an exotic, pleasing hint of sulfur. Hops and malt are not portrayed in the aroma, though there is a suggestion of sweetness and the possibility of floral hops coming through. Light-bodied lager with a barely perceived bitterness. General neutral character, clean, dry finish, with a bit of sweet aftertaste trailing off to nothing. Clean or empty? Take your pick, but it's all that it is supposed to be as an American-style light lager. Refreshing and nonassaultive. A friend of thirst and clean thoughts.

Estimated profile based on tasting
Color: 3 SRM (6 EBC)
Bittering Units: 12–13

Bronze Cup Winner

Labatt Blue
Labatt Breweries of Canada
London, Ontario, Canada

Pale straw color. Clean and extraordinarily neutral aroma with only a slight malt-grain fermented character. Well carbonated. Light in body and exceedingly clean in finish. Refreshes the palate in a neutral manner. Not particularly sweet or bitter. At its own low level, hop bitterness is in balance with malt and adjunct base, skillfully and precisely accomplished. The aftertaste is almost absent. The barest hint of refreshing hop bitterness coupled with a good dose of carbonation indicates this is a refreshing light American lager. It lacks the cloying, sweet aftertaste of many other brands, but the lack of hops still leaves a small residual sweetness that turns to acidity in the aftertaste.

Brewery formulation uses corn as an adjunct.

Estimated profile based on tasting
Original Gravity: 1.044 (11) indicated by the brewery
Final Gravity: 1.006 (1.6) indicated by the brewery
Alcohol by volume: 5% indicated by the brewery
Color: 3–4 SRM (6–8 EBC) (3.2 indicated by the brewery)
Bittering Units: 12–14 (12 indicated by the brewery)

CATEGORY 47:
AMERICAN-STYLE LIGHT LAGER

According to United States' FDA regulations, when the word "light" is used in reference to caloric content, the beer must have at least 25 percent fewer calories than the "regular" version of that beer. Such beers must have certain analysis data printed on the package label. These beers are extremely light-colored, light in body and high in carbonation. Flavor is mild and bitterness is very low. Chill haze, fruity esters and diacetyl should be absent.

Original Gravity (°Plato):	1.024–1.040 (6–10 °Plato)
Apparent Extract- Final Gravity (°Plato):	1.002–1.008 (0.5–2 °Plato)
Alcohol by weight (volume):	2.8–3.5% (3.5–4.4%)
Bitterness (IBU):	8–15
Color (SRM):	2–4 (4–7 EBC)

Gold Cup Winner

Miller Lite
The Miller Brewing Company
3939 W. Highland Blvd.
Milwaukee, Wisconsin, USA 53208
Brewmaster: Dr. David Ryder, vice president, Brewing, Research and Quality Assurance
Established 1855
Production: 45 million bbl. (52.7 million hl.)

Destined to become one of America's most notable German immigrants and brewers, Frederick Miller, born in 1824, began his brewing journey as a wayward scholar-turned-brewing enthusiast. At the age of fourteen he was sent to France for seven years of study, after which he roamed and traveled France, Italy, Switzerland and Algiers. Good fortune led him to visit his uncle's brewery in Nancy, France. He became enraptured, stayed and learned the business and the skills of a brewer. He brewed his first beers at the royal brewery of the Hohenzollerns at Sigmaringen, Germany, which he leased for his own use. Frederick Miller began his career as a contract brewer.

Jeffrey Waalkes

Frederick Miller

In about 1855 Frederick Miller immigrated to the United States and, after working briefly in a New York brewery, found his way to and settled in Milwaukee, which boasted a strong German community. There he bought and reopened the Plank Road Brewery. (See a further account of Miller's origins under the award-winning Icehouse, in the American-style Ice Lager category.) He was on the road to becoming America's second largest producer of beer. Philip Morris bought the Miller Brewing Company in 1969. At the time Miller only had a 4.2 percent share of the American market. Its flagship brand, Miller High Life, led the way, but it wasn't until the successful introduction of Miller Lite that the company's fortunes took off.

Cigars and Miller beer? In the search for a name for their new, light-colored Pilsener lager, the brewery found the answer in the name of a large New Orleans cigar factory. The name: High Life Cigars. The Miller Brewing Company paid $25,000 for the factory and the right to use the name. In

1903 the first bottle of Miller High Life rolled off the bottling lines.

The Miller Brewing Company produces over twenty brands of beer under the main labels of Miller, Sharp's, Milwaukee's Best, Meister Bräu, Löwenbrau, Big Sky and Magnum Malt Liquor. Miller also owns and produces Plank Road Brewery brands, and owns a majority interest in the Jacob Leinenkugel Brewing Co. (Wisconsin), Celis Brewery Inc. (Texas) and the Shipyard Brewery (Maine).

CHARACTER DESCRIPTION OF GOLD CUP–WINNING MILLER LITE

Great tasting. Less filling. It says so right on the bottle. This assessment is quite a change of pace from most of the other beers in this book. Now, where can you go from there? Its color is light gold and as pristine as one could ever imagine. Aroma is reminiscent of a sweet floral hop or pleasant ester. First flavor impression tells it like it is: dry in the palate and very light-bodied, almost like water. A hint of pure pale-malt sweetness, but perhaps the sweet aroma is from corn. Hop bitterness is not perceived but is evident in the smooth overall balance and drinkability. Very clean aftertaste is barely perceptible as bitter. What makes this beer a winner is its exceptionally clean taste, absent of any flavor, good (well, the good ones are at a very low level) or bad. Just what it's supposed to be as a light American beer. It is certainly one of the most challenging beers to duplicate in this book of World Beer Cup winners.

- **Recipe for 5 U.S. gallons (19 liters) Miller Lite**
  Targets:
  Original Gravity: 1.030 (7.5)
  Final Gravity: 1.000 (0)
  Alcohol by volume: 4.2%
  Color: 3 SRM (6 EBC)
  Bittering Units: 16

ALL-GRAIN RECIPE AND PROCEDURE

2	lbs.	(0.9 kg.)	American 2-row pale malt
1¾	lbs.	(0.8 kg.)	American 6-row pale malt
1¾	lbs.	(0.8 kg.)	flaked corn
5½	**lbs.**	**(2.5 kg.)**	**Total grains**

3.5 HBU (99 MBU) American Cluster hops (pellets)—60 minutes (bittering)

¼	tsp.	Irish moss
1	tsp.	amylase enzyme (fungally derived)
¾	c.	corn sugar for priming in bottles. Use ⅓ cup corn sugar if priming a keg.

Wyeast 2007 Pilsen Lager yeast is recommended.

Use very soft, pH-neutral water that is low in minerals (but not absent of calcium). A step infusion mash is employed to mash the grains. Add 5 quarts (4.8 l.) of 130-degree F (54.5 C) water to the crushed grain and flaked corn, stir, stabilize and hold the temperature at 122 degrees F (50 C) for 30 minutes. Add 2.5 quarts (2.4 l.) of boiling water. Add heat if necessary to bring temperature up to 148 degrees F (64.5 C) and hold for about 70 minutes.

After conversion, raise temperature to 167 degrees F (75 C), lauter and sparge with 2.5 gallons (9.5 l.) of 170-degree F (77 C) water. Add additional water to end with 5.5 gallons (21 l.) of sweet liquor. Add bittering hops and bring to a full and vigorous boil.

The total boil time will be 60 minutes. When 10 minutes remain, add Irish moss. After a total wort boil of 60 minutes (reducing the wort volume to just over 5 gallons), turn off the heat, then separate or strain out and sparge hops. Chill the wort to 65 degrees F (18 C) and direct into a sanitized fermenter. Aerate the cooled wort well. Add an active yeast culture and ferment for 4 to 6 days in the primary at 50 degrees F (10 C). Then transfer into a secondary fermenter and add dissolved amylase enzyme. Maintain temperature of 50 degrees F (10 C) to age for two more weeks, then lager for one more week at 40 degrees F (4.5 C).

When secondary aging is complete, prime with sugar, bottle or keg. Let condition at temperatures above 60 degrees F (15.5 C) until clear and carbonated, then store chilled.

MALT-EXTRACT RECIPE AND PROCEDURE FOR MILLER LITE

2¼ lbs. (1 kg.)   English extralight dried malt extract
1½ lbs. (0.7 kg.)   rice extract syrup
0   lb.   (0 kg.)   Total grains

3.5 HBU (99 MBU) American Cluster hops (pellets)—60 minutes (bittering)

¼   tsp.   Irish moss
1   tsp.   amylase enzyme (fungally derived)
¾   c.   corn sugar for priming in bottles. Use ⅓ cup corn sugar if priming a keg.
Wyeast 2007 Pilsen Lager yeast is recommended.

Use very soft, pH-neutral water that is low in minerals. Add extracts and hops to 2.5 gallons (9.5 l.) of water. The total boil time will be 60 minutes. When 10 minutes remain, add Irish moss. After a total wort boil of 60 minutes, turn off the heat, separate or strain out and sparge hops, and direct the hot wort into a sanitized fermenter to which 2 gallons (7.6 l.) of cold water have been added. If necessary, add additional cold water to achieve a 5-gallon (19-l.) batch size. Chill the wort to 70 degrees F (21 C). Aerate the cooled wort well. Add an active yeast culture and ferment for 4 to 6 days in the primary at 50 degrees F (10 C). Then transfer into a secondary fermenter and add dissolved amylase enzyme. Maintain temperature of 50 degrees F (10 C) to age for two more weeks, then lager for one more week at 40 degrees F (4.5 C).

When secondary aging is complete, prime with sugar, bot-

tle or keg. Let condition at temperatures above 60 degrees F (15.5 C) until clear and carbonated, then store chilled.

Silver Cup Winner

Pabst Genuine Draft Light
Pabst Brewing Co.
Milwaukee, Wisconsin, USA

Pale straw color. A neutral aroma with no malt sweetness or hop character. Everything that you expect for this style of beer. Light-bodied with a slightly sweet finish. No bitterness whatsoever. Highly carbonated and clean. A classic character for this style. A great, unmindful beer for times when quenching a thirst is important.

Estimated profile based on tasting
Color: 3–4 SRM (6–8 EBC)
Bittering Units: 10–12

Bronze Cup Winner

Medalla Light Beer
Cerveceria India Ale Inc.
Mayaguez, Puerto Rico

As light in color as a beer can get without being clear. Pristinely clear. Excellent clean, light, sweet, malty character. Exceptionally light-bodied and clean-tasting with a dry, refreshing finish that actually suggests beer. Enough hop bitterness to counterpoint the malt, but not notable as bitterness. Quite assertively carbonated, but balanced with enough flavor to be quite refreshing. Totally neutral in fermented character except for a grainlike base that substantiates the beverage as beer. Aftertaste is vaguely reminiscent of light bitterness that continues as it is enjoyed. Slight herbal-winter-green hop flavor evident during indulgence. Enjoyable at your choice of very cold temperatures or at 50 degrees F.

Estimated profile based on tasting
Original Gravity: 1.032 (8) indicated by the brewery
Final Gravity: 1.001 (0.25) indicated by the brewery
Alcohol by volume: 4.4% indicated by the brewery
Color: 2–3 SRM (4–6 EBC) (3.2 indicated by the brewery)
Bittering Units: 15–18 (13 indicated by the brewery)

CATEGORY 48:
AMERICAN-STYLE PREMIUM LAGER

Similar to the American Lager, this style is a more flavorful, medium-bodied beer and may contain few or no adjuncts at all. Color may be deeper than the American Lager, and alcohol content and bitterness may also be greater. Hop aroma and flavor are low or negligible. Chill haze, fruity esters and diacetyl should be absent. NOTE: Some beers marketed as "premium" (based on price) may not fit this definition.

Original Gravity (°Plato):	1.046–1.050 (11.5–12.5 °Plato)
Apparent Extract–	
Final Gravity (°Plato):	1.010–1.014 (2.5–3.5 °Plato)
Alcohol by weight (volume):	3.6–4% (4.3–5%)
Bitterness (IBU):	13–23
Color (SRM):	2–8 (4–16 EBC)

Gold Cup Winner

Brick Red Baron
Brick Brewing Company
181 King St.
S. Waterloo, Ontario,
   Canada N2J 1P7
Brewmaster: Bill Barnes
Established 1984
Production: 38,500 bbl. (45,000 hl.)

James R.A. Brickman

A brewery that is said to have inspired the microbrewery movement in eastern Canada. Established by entrepreneur, businessman and beer enthusiast Jim Brickman in 1984, the Brick brewery is a testament to resourcefulness, tenacity, quality and vision, all attributes of today's successful microbrewers. Jim Brickman's brewery journey began in 1978. There were only two microbreweries in existence in all of North America. It was a time of skepticism and deaf ears in the banking community, the government and the beer "system" as it existed. Six years later in a 145-year-old abandoned furniture factory, a traditional European brewhouse began producing traditional European-style beers. One of the first breweries in North America to utilize cold sterile filtration to help maintain the freshness of its products, the Brick Brewing Company continues to be a pioneer today. With its focus on specialty products for a small but growing market of beer enthusiasts, the brewery has undergone six expansions since 1984. The success of Jim Brickman and his brewery is testament to the importance of quality. His recipe for success is simple: "You have to be sincere and have a love for brewing." Was he a homebrewer in some former life? Well, at least in spirit, he must have been.

The beers of the Brick Brewing Company include Brick Premium Lager, Brick Red Baron, Brick Amber Dry, Henninger Kaiser Pils (brewed under license from Frankfurt, Germany), Pacific Real Draft, Red Cap Ale, Waterloo Dark and Anniversary Bock.

CHARACTER DESCRIPTION OF GOLD CUP–WINNING BRICK RED BARON

Yellow, golden and pale in color. A perfectly neutral, clean aroma with a very light and insignificant level of green apples. Hop and malt characters are remarkably absent. First flavor impression reveals a medium body, finishing with a pleasant, clean, unpretentious, light pale-malt sweetness. Though low, the malt flavor is ample enough to steer it away from an acidic aftertaste typical of many underhopped American-style light-colored lagers. Hop flavor is negligible, but as with malt character, it is present at levels high enough to contribute to a synergistic overall flavor. Some grainlike adjunct character in nose, appropriate for category.

- ## Recipe for 5 U.S. gallons (19 liters) Brick Red Baron

Targets:
Original Gravity: 1.046 (11.5)
Final Gravity: 1.008 (2)
Alcohol by volume: 5%
Color: 3 SRM (6 EBC)
Bittering Units: 15

ALL-GRAIN RECIPE AND PROCEDURE

8	lbs.	(3.6 kg.)	Canadian 2-row Harrington pale malt
½	lb.	(0.23 kg.)	flaked corn
8½	**lbs.**	**(3.9 kg.)**	**Total grains**

3.2　HBU　(91 MBU) Northern Brewer hops (pellets)—75 minutes (bittering)

1　HBU　(28 MBU) American Hallertauer hops (pellets)—20 minutes (flavor)

¼   tsp.   Irish moss
¾   c.   corn sugar for priming in bottles. Use ⅓ cup
            corn sugar if priming a keg.
Wyeast 2007 Pilsen Lager yeast is recommended.

A step infusion mash is employed to mash the grains. Add 8.5 quarts (8.1 l.) of 130-degree F (54.5 C) water to the crushed grain and flaked corn, stir, stabilize and hold the temperature at 122 degrees F (50 C) for 30 minutes. Add 4 quarts (3.8 l.) of boiling water, adding heat if necessary to bring temperature up to 150 degrees F (65.5 C). Hold for about 60 minutes.

After conversion, raise temperature to 167 degrees F (75 C), lauter and sparge with 4 gallons (15 l.) of 170-degree F (77 C) water. Collect about 6 gallons (23 l.) of runoff, add bittering hops and bring to a full and vigorous boil.

The total boil time will be 75 minutes. When 20 minutes remain, add flavor hops. When 10 minutes remain, add Irish moss. After a total wort boil of 75 minutes (reducing the wort volume to just over 5 gallons), turn off the heat, add the aroma hops and let steep for 3 to 5 minutes, then separate or strain out and sparge hops. Chill the wort to 65 degrees F (18 C) and direct into a sanitized fermenter. Aerate the cooled wort well. Add an active yeast culture and ferment for 4 to 6 days in the primary at 50 degrees F (10 C). Then transfer into a secondary fermenter. Maintain temperature of 50 degrees F (10 C) to age for two more weeks, then lager for two to four more weeks at 40 degrees F (4.5 C).

When secondary aging is complete, prime with sugar, bottle or keg. Let condition at temperatures above 60 degrees F (15.5 C) until clear and carbonated, then store chilled.

MASH-EXTRACT RECIPE AND PROCEDURE FOR BRICK RED BARON

4	lbs.	(1.8 kg.)	English extralight dried malt extract
1½	lbs.	(0.7 kg.)	Canadian Harrington 2-row pale malt
½	lb.	(0.7 kg.)	flaked corn
2	**lbs.**	**(0.9 kg.)**	**Total grains**

3.6 HBU (102 MBU) Northern Brewer hops (pellets)—
     60 minutes (bittering)

1   HBU (28 MBU) American Hallertauer hops (pel-
     lets)—20 minutes (flavor)

¼   tsp. Irish moss
¾   c.   corn sugar for priming in bottles. Use ⅓ cup
           corn sugar if priming a keg.
Wyeast 2007 Pilsen Lager yeast is recommended.

A step infusion mash is employed to mash the grains.
Add 2 quarts (1.9 l.) of 130-degree F (54.5 C) water to the
crushed grain and flaked corn, stir, stabilize and hold the
temperature at 122 degrees F (50 C) for 30 minutes. Add 1
quart (1 l.) of boiling water, add heat to bring temperature
up to 150 degrees F (65.5 C), and hold for about 60 minutes.

After conversion, raise temperature to 167 degrees F (75
C), lauter and sparge with 1 gallon (3.8 l.) of 170-degree F
(77 C) water. Collect about 1.5 gallons (5.7 l.) of runoff. Add
additional 1 gallon (3.8 l.) of water and the malt extract and
bittering hops. Bring the 2.5 gallons (9.5 l.) to a full and
vigorous boil.

The total boil time will be 60 minutes. When 20 minutes
remain, add flavor hops. When 10 minutes remain, add Irish
moss. After a total wort boil of 60 minutes, turn off the heat,
then separate or strain out and sparge hops, and direct the
hot wort into a sanitized fermenter to which 2 gallons (7.6
l.) of cold water have been added. If necessary, add additional cold water to achieve a 5-gallon (19-l.) batch size.
Chill the wort to 65 degrees F (18 C) and direct into a
sanitized fermenter. Aerate the cooled wort well. Add an
active yeast culture and ferment for 4 to 6 days in the primary at 50 degrees F (10 C). Then transfer into a secondary
fermenter. Maintain temperature of 50 degrees F (10 C) to
age for two more weeks, then lager for two to four more
weeks at 40 degrees F (4.5 C).

When secondary aging is complete, prime with sugar, bot-

tle or keg. Let condition at temperatures above 60 degrees F (15.5 C) until clear and carbonated, then store chilled.

Silver Cup Winner

Signature
Stroh Brewery Co.
Detroit, Michigan, USA

Pale golden color. Light-hearted hop aroma attempts to emerge beyond a subdued grainy malt aroma. Mouth feel is light-bodied with a dry finish. As a premium style, it is distinctive for having subdued grain and hop characters. Extremely low level of DMS (sweet corn character) helps define the flavor of this lager. A memory of malt sweetness and a soft hop bitterness. The hopping rate is skillfully accomplished to balance the lighter characters of this American-style premium lager. At second impression, the body moves toward the medium range, perhaps because of the evident hops and the sweet cornlike, grainy flavor not evident in comparatively lighter styles of American lager.

Brewery formulation uses corn adjunct with Czech Saaz and Slovenian Styrian Goldings for bitterness, flavor and aroma.

Estimated profile based on tasting
Original Gravity: 1.048 (12) indicated by the brewery
Final Gravity: 1.018 (4.5) indicated by the brewery
Alcohol by volume: 4.9% indicated by the brewery
Color: 5-6 SRM (10–12 EBC) (4.6 SRM indicated by the brewery)
Bittering Units: 15–17 (16 indicated by the brewery)

Bronze Cup Winner

Budweiser
Anheuser-Busch Inc.
St. Louis, Missouri, USA

Extremely pale in color. Very subtle applelike fruitiness is evident in the aroma. Sweet malt aroma also emerges, but hop character is absent. Bitterness is negligible and only present to balance out malt sweetness. A light-bodied beer with a dry finish. Fully and notably carbonated. Aftertaste is primarily neutral with a faint memory of malt. A classic American-style lager. Refreshing for its mildness. Other than drinkability, memorableness is not particularly important for this style and is not evident in this beer. A quintessential example of the American premium lager style.

Estimated profile based on tasting
Color: 3–4 SRM (6–8 EBC)
Bittering Units: 12–13

CATEGORY 49: DRY LAGER

This straw-colored lager lacks sweetness and is reminiscent of an American-style light lager. However, its starting gravity and alcoholic strength are greater. Hop rates are low and carbonation is high. Chill haze, fruity esters and diacetyl should be absent.

Original Gravity (°Plato):	1.040–1.050 (10–12.5 °Plato)
Apparent Extract–	
Final Gravity (°Plato):	1.004–1.008 (1–2 °Plato)
Alcohol by weight (volume):	3.6–4.5% (4.3–5.5%)
Bitterness (IBU):	15–23
Color (SRM):	2–4 (4–7 EBC)

Gold Cup Winner

None

Silver Cup Winner

None

Bronze Cup Winner

Cerveja Antarctica
Cia. Antarctica Paulista
Jaguariúna, São Paulo, Brazil

Pale straw color. Clean, grainy and sweet malt aroma emerges delicately along with a hint of pleasantly sweet, smoky phenols mixed with a low and pleasant level of corn (DMS). Very evocative aroma reminiscent of Brazil's lifestyle; a flavor with subtle passions that intensify with each sip. Pleasant bitterness evolves with each taste rather than diminishing. Light-bodied mouth feel with a dry finish. Not excessively sweet. Middle experience provides a nice, refreshing bitter bite that is neither harsh nor soft. A roundness of flavor portrays the complexity of the malt and grain flavors that continue through the experience, without being assertive. Refreshing and clean. Hop flavor and aroma absent.

Estimated profile based on tasting
Color: 3–4 SRM (6–8 EBC)
Bittering Units: 19–23

CATEGORY 50:
AMERICAN-STYLE ICE LAGER

This style is slightly higher in alcohol than most other light-colored, American-style lagers. Its body is low to medium and it has a low residual malt sweetness. It has few or no adjuncts. Color is very pale to golden. Hop bitterness is low, but is certainly perceptible. Hop aroma and flavor are low. Chill haze, fruity esters and diacetyl should not be perceived. Typically these beers are chilled before filtration so that ice crystals (which may or may not be removed) are formed. This can contribute to a higher alcohol content (up to 0.5 percent more).

Original Gravity (°Plato):	1.040–1.060 (10–15 °Plato)
Apparent Extract– Final Gravity (°Plato):	1.006–1.014 (1.5–3.5 °Plato)
Alcohol by weight (volume):	3.8–5% (4.6–6%)
Bitterness (IBU):	7–20
Color (SRM):	2–8 (5–16 EBC)

Gold Cup Winner

Icehouse
Plank Road Brewery
4400 W. State Street
Milwaukee, Wisconsin, USA 53201
Brewmaster: Dr. David Ryder, vice president, Brewing Research and Quality Assurance
Established 1993
Production: 3 million bbl. (3.5 million hl.)

Plank Road Brewery, a division of Miller Brewing Company, gets its name from the location and original name of the

Jeffrey Waalkes

The Plank-Road Brewery

first brewery founded in Milwaukee by Frederick Miller in the 1850s. In 1873 the Plank Road Brewery changed its name to the Menomonee Valley Brewery, eventually becoming the Miller Brewing Company.

The Plank Road Brewery was envisioned by the Miller Brewing Company as having an entrepreneurial spirit in keeping with the small, handcrafted microbreweries of the Northwest. In 1993 the Miller Brewing Company saw an opportunity to tap into the growing specialty-beer market. Though the market is small, Miller believed that some mainstream beer drinkers wanted more than the average American mass-produced beer. Plank Road philosophy takes the spirit of Miller back to the simpler times of a small brewery by creating and marketing a high-quality brew for the changing tastes of many of today's mainstream beer drinkers. By allowing a group of dedicated Miller employees to launch their "own" brewing division, Miller shed some of its corporate bureaucracy and gave a select group of employees an opportunity to become more creative. Today Plank Road products are targeted at the average American beer drinker who prefers beers whose image is positioned in the marketplace between the lighter-flavored mainstream beers and the full-flavored microbrewed beers.

Plank Road products include the award-winning Icehouse and Red Dog, Southpaw Light and Northstone Amber Ale.

CHARACTER DESCRIPTION OF GOLD CUP–WINNING ICEHOUSE

"Ice brewed to eliminate watered-down taste." That's what it says on the label. Pale gold color. Aroma is slightly sweet with a minuscule level of sweet cornlike lager character. Mouth feel presents itself as low approaching a medium body. Flavor is a simple fermented malt character. No flaws. No bitterness except to balance the aftertaste and effectively neutralize flavor. Other than a hint of sulfur-yeast character, Icehouse is extraordinarily absent of fermentation flavor character.

- **Recipe for 5 U.S. gallons (19 liters) Icehouse**
  Targets:
  Original Gravity: 1.045 (11.25)
  Final Gravity: 1.006 (1.5)
  Alcohol by volume: 5.2%
  Color: 3.8 SRM (7.5 EBC)
  Bittering Units: 15

ALL-GRAIN RECIPE AND PROCEDURE

4¼ lbs.	(1.9 kg.)	American 6-row pale malt
4 lbs.	(1.8 kg.)	flaked corn
0.1 oz.	(4 g.)	American black malt
8¼ lbs.	**(3.7 kg.)**	**Total grains**

| 2 | HBU | (56 MBU) American Cluster hops (pellets)—75 minutes (bittering) |
| 1 | HBU | (28 MBU) American Galena hops (pellets)—75 minutes (bittering) |

| ¼ | tsp. | Irish moss |
| ¾ | c. | corn sugar for priming in bottles. Use ⅓ cup corn sugar if priming a keg. |

Wyeast 2007 Pilsen Lager yeast is recommended.

A step infusion mash is employed to mash the grains. Add 8.5 quarts (8.1 l.) of 130-degree F (54.5 C) water to the crushed grain and flaked corn, stir, stabilize and hold the temperature at 122 degrees F (50 C) for 30 minutes. Add 4 quarts (3.8 l.) of boiling water, adding heat if necessary to bring temperature up to 150 degrees F (65.5 C). Hold for about 60 minutes.

After conversion, raise temperature to 167 degrees F (75 C), lauter and sparge with 4 gallons (15 l.) of 170-degree F (77 C) water. Collect about 6 gallons (23 l.) of runoff, add bittering hops and bring to a full and vigorous boil.

The total boil time will be 75 minutes. When 10 minutes remain, add Irish moss. After a total wort boil of 75 minutes (reducing the wort volume to just over 5 gallons), turn off the heat, then separate or strain out and sparge hops. Chill the wort to 65 degrees F (18 C) and direct into a sanitized fermenter. Aerate the cooled wort well. Add an active yeast culture and ferment for 4 to 6 days in the primary at 50 degrees F (10 C), then transfer into a secondary fermenter. Maintain temperature of 50 degrees F (10 C) to age for two more weeks, then lager for two to four more weeks at 40 degrees F (4.5 C). After lagering, drop temperature to 30 to 31 degrees F (-1 C) to encourage ice in the form of a slight slush on the surface. Separate beer from ice slush.

Prime with sugar then bottle or keg. Let condition at temperatures above 60 degrees F (15.5 C) until clear and carbonated, then store chilled.

MASH-EXTRACT RECIPE AND PROCEDURE FOR ICEHOUSE

2¼	lbs.	(1 kg.)	English light dried malt extract
2½	lbs.	(1.1 kg.)	American 6-row pale malt
2	lbs.	(0.9 kg.)	flaked corn
0.1	oz.	(4 g.)	American black malt
4½	**lbs.**	**(2 kg.)**	**Total grains**

2   HBU (56 MBU) American Cluster hops (pellets)—75 minutes (bittering)

2   HBU (56 MBU) American Galena hops (pellets)—75 minutes (bittering)

¼   tsp.   Irish moss

¾   c.   corn sugar for priming in bottles. Use ⅓ cup corn sugar if priming a keg.

Wyeast 2007 Pilsen Lager yeast is recommended.

A step infusion mash is employed to mash the grains. Add 4.5 quarts (4.3 l.) of 130-degree F (54.5 C) water to the crushed grain and flaked corn, stir, stabilize and hold the temperature at 122 degrees F (50 C) for 30 minutes. Add 2 quarts (1.9 l.) of boiling water. Add heat to bring temperature up to 150 degrees F (65.5 C). Hold for about 60 minutes.

After conversion, raise temperature to 167 degrees F (75 C), lauter and sparge with 2 gallons (3.8 l.) of 170-degree F (77 C) water. Collect about 3 gallons (11.4 l.) of runoff. Add malt extract and bittering hops and bring to a full and vigorous boil.

The total boil time will be 75 minutes. When 10 minutes remain, add Irish moss. After a total wort boil of 75 minutes (reducing the wort volume to just over 5 gallons), turn off the heat, then separate or strain out and sparge hops, and direct the hot wort into a sanitized fermenter to which 2 gallons (7.6 l.) of cold water have been added. If necessary, add additional cold water to achieve a 5-gallon (19-l.) batch size. Chill the wort to 65 degrees F (68.5 C). Aerate the cooled wort well. Add an active yeast culture and ferment for 4 to 6 days in the primary at 50 degrees F (10 C). Then transfer into a secondary fermenter. Maintain temperature of 50 degrees F (10 C) to age for two more weeks, then lager for two more weeks at 40 degrees F (4.5 C). After lagering, drop temperature to 30 to 31 degrees F (-1 C) to encourage ice in the form of a slight slush on the surface. Separate beer from ice slush.

Prime with sugar then bottle or keg. Let condition at temperatures above 60 degrees F (15.5 C) until clear and carbonated, then store chilled.

Silver Cup Winner

Molson Ice
Molson Breweries-MCI
Etobicoke, Ontario, Canada

Pale straw color. Sweet malt aroma emerges with a hint of floral character. Fresh, clean beer character. Medium-bodied lager. First flavor impression portrays sweetness with overall neutrality. Very smooth, with bitterness hardly perceptible. Hop flavor and aroma absent. Malt sweetness and skillfully lagered sweet esters are very smooth and pleasant. A sweet aftertaste that gradually dissipates to a neutral, slightly alcoholic flavor. The label indicates that the beer "is slow brewed and super chilled until ice crystals form, then filtered and blended with pure Canadian water."

Estimated profile based on tasting
Alcohol by volume: 5.6% as indicated on label
Color: 4–5 SRM (8–10 EBC)
Bittering Units: 12–14

Bronze Cup Winner

Schlitz Ice
Jos. Schlitz Brewing Co.
Detroit, Michigan, USA

Pale golden in color. Sweet, very clean base malt-and-grain aroma with no evidence of higher-type alcohols. Flavor is medium-bodied followed by a dry finish. Strinkingly sweet flavor finishes off with a small bite of bitterness and warming alcohol. This beer is so dry and light, the flavor of the alcohol is certainly evident, becoming a part of the overall flavor profile. Aftertaste is a memorable hit of alcohol flavor, but is very clean with a lingering sweetness and a continuing nip of bitterness. Clean, simple and strong.

Estimated profile based on tasting
Original Gravity: 1.044 (11) indicated by the brewery
Final Gravity: 1.008 (4) indicated by the brewery
Alcohol by volume: 4.6% indicated by the brewery, though label says 7.7%
Color: 4 SRM (8 EBC) (3.2 indicated by the brewery)
Bittering Units: 14–16 (14 indicated by the brewery)

CATEGORY 51: AMERICAN-STYLE MALT LIQUOR

High in starting gravity and alcoholic strength, this style is somewhat diverse. Some American Malt Liquors are just slightly stronger than American lagers, while others approach bock strength. Some residual sweetness is perceived. Hop rates are very low, contributing little bitterness and virtually no hop aroma or flavor. Chill haze, diacetyl and fruity esters should not be perceived.

Orginal Gravity (°Plato):	1.050–1.060 (12.5–15 °Plato)
Apparent Extract– Final Gravity (°Plato):	1.004–1.010 (1–2.5 °Plato)
Alcohol by weight (volume):	5–6% (6.25–7.5%)
Bitterness (IBU):	12–23
Color (SRM):	2–5 (4–8 EBC)

Gold Cup Winner

Olde English 800 Malt Liquor
Pabst Brewing Company
622 East Vienna Ave.
Milwaukee, Wisconsin, USA 53212
Brewmaster: Bob Newman
Established 1844 as Best and Company
Production: 6 million bbl. (7 million hl.)

The original brewkettle had a capacity of just 18 barrels. That's a microbrewery by anyone's standard. By 1893 the brewery had expanded and had changed its name from Best and Company to Phillip Best Brewing Company. It eventually became the Pabst Brewing Company after Captain Fred Pabst married into the company. He led the brewery to become the largest in the United States at one million barrels of production a year. A hundred years ago these were very large numbers.

Pabst Brewery claims to have brewed the first lager beer in Milwaukee. But in actuality there had to have been homebrewers who perhaps had been meeting their own needs for lager. No one knows for sure, but one just can't help but imagine that the thirst for beer hasn't changed over the centuries.

Joyce Talatzko

Malt liquor is usually much maligned by beer and home-brew enthusiasts, but they might be interested in hearing a story about how I prove to some very knowledgeable beer judges that we are indeed prejudiced by marketing and image. In a blind beer tasting I once slipped a bottle of malt liquor onto the palates of several beer enthusiasts, claiming that it was a Maibock. The assessment was that the brew lacked a bit of body and malt character but that it seemed a plausible example of this traditional German style of lager. The panel was flabbergasted when told what the beer actually was.

Olde English 800 Malt Liquor is a proven winner of this category in the World Beer Cup. Different panels of judges have consistently awarded Olde English 800 Gold Medals at the 1991, 1992, 1994 and 1995 Great American Beer Festivals. This is no small feat as the ever-more-sophisticated palates of judges from around the world assess these beers.

Beers produced by the Pabst Brewing Company include a range of products under the brands of Pabst, Olympia, Andeker, Jacob Best and Hamm's.

CHARACTER DESCRIPTION OF GOLD CUP–WINNING OLDE ENGLISH 800 MALT LIQUOR

Golden yellow color. A touch of berrylike fruitiness, sweet honeylike malt and alcohol aroma greets one upon first indulgence. Immediate impression of mouth feel is barely beyond low in body, followed by a very dry after-taste-finish. Overall Olde English 800 flavor reveals a sweetness that is not heavy and only enough bitterness to minimally balance the sweet character. A mild-tasting beer with no strong flavors and very little aftertaste. With further indulgence one begins to perceive a tiny bit of bitterness in aftertaste, but then again, this may be an illusionary effect of alcohol.

- **Recipe for 5 U.S. gallons (19 liters) Olde English 800 Malt Liquor**
  Targets:
  Original Gravity: 1.055 (13.5)
  Final Gravity: 1.004 (1)
  Alcohol by volume: 7%
  Color: 4.5 SRM (9 EBC)
  Bittering Units: 14

ALL-GRAIN RECIPE AND PROCEDURE

3½ lbs.	(1.6 kg.)	American 2-row pale malt
3¾ lbs.	(1.7 kg.)	American 6-row pale malt
3 lbs.	(1.4 kg.)	flaked corn
10¼ lbs.	**(4.7 kg.)**	**Total grains**

2	HBU	(56 MBU) American Cluster hops (pellets)—105 minutes (bittering)
1.5	HBU	(43 MBU) American Nugget hops (pellets)—105 minutes (bittering)

¼	tsp.	Irish moss
¾	c.	corn sugar for priming in bottles. Use ⅓ corn sugar if priming a keg.

Wyeast 2007 Pilsen Lager yeast is recommended.

A step infusion mash is employed to mash the grains. Add 10 quarts (9.5 l.) of 130-degree F (54.5 C) water to the crushed grain and flaked corn, stir, stabilize and hold the temperature at 122 degrees F (50 C) for 30 minutes. Add 5 quarts (1.9 l.) of boiling water. Add heat to bring temperature up to 150 degrees F (65.5 C). Hold for about 60 minutes.

After conversion, raise temperature to 167 degrees F (75 C), lauter and sparge with 4 gallons (15 l.) of 170-degree F (77 C) water. Collect about 6.5 gallons (25 l.) of runoff, add bittering hops and bring to a full and vigorous boil.

The total boil time will be 105 minutes. When 10 minutes remain, add Irish moss. After a total wort boil of 105 minutes (reducing the wort volume to just over 5 gallons), turn off the heat, then separate or strain out and sparge hops. Chill the wort to 65 degrees F (18 C) and direct into a sanitized fermenter. Aerate the cooled wort well. Add an active yeast culture and ferment for 4 to 6 days in the primary at 55 degrees F (15 C). Then transfer into a secondary fermenter, chill to 50 degrees F (10 C) to age for two more weeks, then lager for two to four more weeks at 40 degrees F (4.5 C).

When secondary aging is complete, prime with sugar, bottle or keg. Let condition at temperatures above 60 degrees F (15.5 C) until clear and carbonated, then store chilled.

MASH-EXTRACT RECIPE AND PROCEDURE FOR OLDE ENGLISH 800 MALT LIQUOR

2½	lbs. (1.1 kg.)	English light dried malt extract
3½	lbs. (1.6 kg.)	American 6-row pale malt
2½	lbs. (1.1 kg.)	flaked corn
6	**lbs. (2.7 kg.)**	**Total grains**

2	HBU	(56 MBU) American Cluster hops (pellets)—75 minutes (bittering)
2	HBU	(56 MBU) American Nugget hops (pellets)—75 minutes (bittering)

¼	tsp.	Irish moss
¾	c.	corn sugar for priming in bottles. Use ⅓ cup corn sugar if priming a keg.

Wyeast 2007 Pilsen Lager yeast is recommended.

A step infusion mash is employed to mash the grains. Add 6 quarts (5.7 l.) of 130-degree F (54.5 C) water to the crushed grain and flaked corn, stir, stabilize and hold the temperature at 122 degrees F (50 C) for 30 minutes. Add 3 quarts (2.9 l.) of boiling water, add heat to bring temperature up to 150 degrees F (65.5 C), and hold for about 60 minutes.

After conversion, raise temperature to 167 degrees F (75 C), lauter and sparge with 2 gallons (7.6 l.) of 170-degree F (77 C) water. Collect about 3 gallons (11.5 l.) of runoff. Add malt extract and bittering hops and bring to a full and vigorous boil.

The total boil time will be 75 minutes. When 10 minutes remain, add Irish moss. After a total wort boil of 75 minutes, turn off the heat, then separate or strain out and sparge hops, and direct the hot wort into a sanitized fermenter to which 2 gallons (7.6 l.) of cold water have been added. If necessary, add additional cold water to achieve a 5-gallon (19-l.) batch size. Chill the wort to 70 degrees F (21 C). Aerate the cooled wort well. Add an active yeast culture and ferment for 4 to 6 days in the primary. Then transfer into a secondary fermenter and chill to 55 to 60 degrees F (13–15.5 C) if possible. Allow to age for two weeks or more.

When secondary aging is complete, prime with sugar, bottle or keg. Let condition at temperatures above 60 degrees F (15.5 C) until clear and carbonated.

Silver Cup Winner

Schlitz Malt Liquor
Jos. Schlitz Brewing Co.
Detroit, Michigan, USA

Pale gold in color. Neither hops nor malt emerges in the very neutrally clean aroma. Second aromatic impression portrays some graininess. Flavor has a dry finish contributed to by high carbonation. A hint of fruitiness complements the flavor along with the alcohol's tingling sensation and flavor. A small dose of hop bitterness synergizes with the alcohol to create an overall balanced neutral, quenching, dry beer. Aftertaste is mildly suggestive of refreshing hop bitterness. Extraordinarily clean, with an expected hint of high-alcohol fruitiness.

Brewery formulation uses corn adjunct with American Galena hops for bitterness.

Estimated profile based on tasting
Original Gravity: 1.054 (13.4) indicated by the brewery
Final Gravity: 1.018 (4.5) indicated by the brewery
Alcohol by volume: 5.9% indicated by the brewery
Color: 4–5 SRM (8–10 EBC) (4.6 SRM indicated by the brewery)
Bittering Units: 13-16 (12 indicated by the brewery)

Bronze Cup Winner

Country Club Malt Liquor
Pearl Brewing Co.
San Antonio, Texas, USA

Straw golden in color. No hop aroma. Sweetness borrows a good dose of neutral adjunct character from corn or other sugars. A small hint of DMS (cornlike character) is evident. There is a quiet sense of alcohol in the aroma while most other aromatic characters are minimal or neutral. Mouth feel is medium-bodied with the a light aftertaste. Some bitterness in aftertaste, which plays an essential role in balancing the sweetness of character. Alcohol has some warming effect in flavor and aftertaste, though overall this is a simple, straightforward beer for malt liquor connoisseurs.

Estimated profile based on tasting
Color: 3–4 SRM (6–8 EBC)
Bittering Units: 15–17

CATEGORY 52:
AMERICAN-STYLE AMBER LAGER

American-Style Amber Lagers are amber, reddish brown or copper-colored. They are medium-bodied. There is a noticeable degree of caramel-type malt character in flavor and often in aroma. This is a broad category in which the hop bitterness, flavor and aroma may be accentuated or at relatively low levels, yet noticeable. Fruity esters, diacetyl and chill haze should be absent.

Original Gravity (°Plato):   1.042–1.056 (10.5–14 °Plato)
Apparent Extract–
Final Gravity (°Plato):   1.010–1.018 (2.5–4.5 °Plato)
Alcohol by weight (volume):   3.8–4.3% (4.8–5.4%)
Bitterness (IBU):   20–30
Color SRM (EBC):   6–12 (15–30 EBC)

Gold Cup Winner

Point Amber Lager
Stevens Point Brewery
2617 Water St.
Stevens Point, Wisconsin, USA 54481
Brewmaster: John M. Zappa
Established 1857
Production: 49,000 bbl. (57,000 hl.)

I'll never forget a personal visit I made to the Stevens Point Brewery in 1988. "Astounding," I thought. "A microbrewery that has survived in America for 131 years." Since 1857 this small local brewery has survived the Civil War, two World Wars, the Great Depression, American Prohibition and the competition of larger breweries. It is one of only sixteen

John Zappa, Brewmaster

continuously operating pre-Prohibition brewing companies in the United States. No small feat in a country where there were thousands of breweries across the land at the turn of the twentieth century. Interestingly, Stevens Point Brewery was actually founded a year earlier than the city of Stevens Point. Ah. It's amazing what a good little brewery can inspire.

I recall that in 1988 Stevens Point was an immaculately well-kept small brewery in the heartland of America, barely surviving at a production level of about 20,000 barrels, a dream production for most aspiring microbreweries of the day. But at the time the Stevens Point Brewery produced primarily American-style light lagers, competing little-head-to-big-head with the national brands and slowly losing out. Fortunately for beer enthusiasts, this proud little brewery adapted to market changes in time to produce the quality beer sought by America's new beer enthusiasts.

I can only let my imagination drift upward and beyond along with the surfacing bubbles of award-winning Point

Classic Amber and believe that when Frank Wahle and George Ruder established the Stevens Point Brewery in the early days of lumberjacks and woodlands, they first had been homebrewers with a dream and a full-flavored thirst. They just had to be, didn't they?

Other beers produced by the Stevens Point Brewery are Point Special, Point Bock, Point Pale Ale, Point Maple Wheat and Point Winter Spice.

CHARACTER DESCRIPTION OF GOLD CUP–WINNING POINT AMBER LAGER

Tawny amber color with a reddish hue. Promoted as a 100 percent barley "Point Classic" and brewed with three hops, it has a firm but subtle caramel character and an all-malt aroma. Hop aroma is subtle and gentle, and expresses the soft sweetness of American Tettnanger hops. Bitterness is not part of a first flavor impression, but becomes apparent as a combination of mid- to low-alpha American hops. Body is low to medium in mouth feel, with an aftertaste that includes the medium intensity of sweet caramel malt. Bitterness is overridden by dominant, though gentle, malt character. Extraordinarily clean beer absent of complex fermentation byproducts.

• **Recipe for 5 U.S. gallons (19 liters) Point Amber Lager**

Targets:

Original Gravity: 1.048 (12)

Final Gravity: 1.012 (3)

Alcohol by volume: 4.8%

Color: 16 SRM (32 EBC)

Bittering Units: 15

ALL-GRAIN RECIPE AND PROCEDURE

4	lbs. (1.8 kg.)	American 2-row pale malt
3¾	lbs. (1.7 kg.)	American 6-row pale malt

1¼	lbs.	(0.6 kg.)	American caramel malt—60 Lovibond
¼	lb.	(114 g.)	American Cara-Pils malt
9¼	**lbs.**	**(4.2 kg.)**	**Total grains**

2	HBU	(56 MBU) American Cluster hops (pellets)—90 minutes (bittering)
1	HBU	(28 MBU) American Hallertauer hops (pellets)—30 minutes (flavor)
1	HBU	(28 MBU) American Tettnanger hops (pellets)—30 minutes (flavor)
¼	oz.	(7 g.) American Cascade hops (pellets)—steep in finished boiled wort for 2 to 3 minutes (aroma)

| ¼ | tsp. | Irish moss |
| ¾ | c. | corn sugar for priming in bottles. Use ⅓ cup corn sugar if priming a keg. |

Wyeast 2007 Pilsen Lager yeast

A step infusion mash is employed to mash the grains. Add 9 quarts (8.5 l.) of 140-degree F (60 C) water to the crushed grain, stir, stabilize and hold the temperature at 133 degrees F (56 C) for 30 minutes. Add 4.5 quarts (4.3 l.) of boiling water, add heat if necessary to bring temperature up to 153 degrees F (67 C), and hold for about 60 minutes.

After conversion, raise temperature to 167 degrees F (75 C), lauter and sparge with 4 gallons (15 l.) of 170-degree F (77 C) water. Collect about 6 gallons (23 l.) of runoff, add bittering hops and bring to a full and vigorous boil.

The total boil time will be 90 minutes. When 30 minutes remain, add flavor hops. When 10 minutes remain, add Irish moss. After a total wort boil of 90 minutes (reducing the wort volume to just over 5 gallons), turn off the heat, add aroma hops and let steep for 3 to 5 minutes. Then separate or strain out and sparge hops. Chill the wort to 65 degrees F (18 C) and direct into a sanitized fermenter. Aerate the cooled wort well. Add an active yeast culture and ferment for 4 to 6 days in the primary at 55 degrees F (15 C). Then transfer into a secondary fermenter, chill to 50 degrees F (10 C) to age for two more weeks, then lager for two more weeks at 40 degrees F (4.5 C).

When secondary aging is complete, prime with sugar, bottle or keg. Let condition at temperatures above 60 degrees F (15.5 C) until clear and carbonated, then store chilled.

MALT-EXTRACT RECIPE AND PROCEDURE FOR POINT AMBER LAGER

4¾ lbs. (2.2 kg.)   English light dried malt extract
1¼ lbs. (0.6 kg.)   American caramel malt—60 Lovibond
1¼ lbs. (0.57 kg) Total grains

3 HBU (85 MBU) American Cluster hops (pellets)—60 minutes (bittering)
1 HBU (28 MBU) American Hallertauer hops (pellets)—30 minutes (flavor)
1 HBU (28 MBU) American Tettnanger hops (pellets)—30 minutes (flavor)
¼ oz. (7 g.) American Cascade hops (pellets)—steep in finished boiled wort for 2 to 3 minutes (aroma)

¼ tsp. Irish moss
¾ c. corn sugar for priming in bottles. Use ⅓ cup corn sugar if priming a keg.
Wyeast 2007 Pilsen Lager yeast

Steep crushed specialty grains in 1½ gallons (5.7 l.) water at 150 degrees F (65.5 C) for 30 minutes. Strain and sparge with enough 170-degree F (76.5 C) water to finish with a little over 2½ gallons (9.5 l.) specialty grain liquor. Add the dried malt extract and bittering hops and bring to a full and vigorous boil.

The total boil time will be 60 minutes. When 10 minutes remain, add Irish moss. After a total wort boil of 60 minutes, turn off the heat, separate or strain out and sparge hops, and direct the hot wort into a sanitized fermenter to which 2 gallons (7.6 l.) of cold water have been added. If necessary, add additional cold water to achieve a 5-gallon (19-l.) batch size. Aerate the cooled wort well. Add an active yeast culture and ferment for 4 to 6 days in the primary at 55 degrees F (15 C). Then transfer into a secondary fermenter, chill to 50 degrees F (10 C) to age for two more weeks, then lager for two more weeks at 40 degrees F (4.5 C).

When secondary aging is complete, prime with sugar, bottle or keg. Let condition at temperatures above 60 degrees F (15.5 C) until clear and carbonated, then store chilled.

Silver Cup Winner

JJ Wainwright Evil Eye Amber Lager
Pittsburgh Brewing Co.
Pittsburgh, Pennsylvania, USA

Light amber color with orange hue. Complex American-type hop aroma (such as citrusy Cascade) emerges. Malt sweetness is evident, but hop aroma predominates. Light to medium body with a clean, refreshing, neutral flavor. A mild Munich/Vienna malt flavor accompanied by a very light caramel character balances a reflective bitterness that emerges in the aftertaste. Hop flavor is generally fruity and citrusy but not as strong as in aroma. Overall impression returns to hop character with mild malt character.

Estimated profile based on tasting
Color: 9–11 SRM (18–22 EBC)
Bittering Units: 21–27

Bronze Cup Winner

Red Wolf
Anheuser-Busch Inc.
St. Louis, Missouri, USA

Color is midway between amber and brown with orange hues. Relatively neutral aroma with a slight suggestion of caramel that stops short of making a definitive statement. Flavor offers a sweet impression with clean maltiness. A pleasant small bitterness creeps into the finish. Medium-bodied mouth feel, though the finish leaves a light impression. Toasted malts contribute mostly color and some roast bitterness, but overall do not contribute significantly to flavor.

Estimated profile based on tasting
Alcohol by volume: 5.5% as indicated on label
Color: 12–13 SRM (24–26 EBC)
Bittering Units: 18–20

CATEGORY 53:
AMERICAN-STYLE DARK LAGER

This beer's maltiness is less pronounced than European dark lagers, and the body is light. Nonmalt adjuncts are often used, and hop rates are low. Hop bitterness flavor and aroma are low. Carbonation is high and is more typical of an American-style light lager than a European dark lager. Fruity esters, diacetyl and chill haze should not be perceived.

Original Gravity (°Plato): 1.040–1.050 (10–12.5 °Plato)
Apparent Extract–
Final Gravity (°Plato): 1.008–1.012 (2–3 °Plato)

Alcohol by weight (volume):   3.2–4.4% (4–5.5%)
Bitterness (IBU):   14–20
Color (SRM):   10–20 (25–80 EBC)

Gold Cup Winner

None

Silver Cup Winner

None

Bronze Cup Winner

None

Other Origin

CATEGORY 54:
TROPICAL-STYLE LIGHT LAGER

Very light in color. Light-bodied. No hop flavor or aroma. Hop bitterness is negligible to moderately perceived. Sugar adjuncts are often used to lighten the body and flavor, sometimes contributing to a slight applelike fruity ester. Sugar, corn, rice and other cereal grains are used as adjuncts. Chill haze and diacetyl should be absent. Fruity esters should be very low.

Original Gravity (°Plato):   1.032–1.046 (8–11.5 °Plato)
Apparent Extract–
Final Gravity (°Plato):   1.004–1.010 (1–2.5 °Plato)
Alcohol by weight (volume):   2.0–4.5% (2.5–5.6%)
Bitterness (IBU):   9–25
Color (SRM):   2–4 (6–10 EBC)

Gold Cup Winner

Cascade Pale Ale
The Cascade Brewery Co.
Cascades
Hobart, Tasmania, Australia
Brewmaster: Frank Messina
Established 1832
Production: 171,000 bbl. (200,000 hl.)

Designed and planned in the confines of a Tasmanian prison cell, the Cascade Brewery has emerged as one of Australia's brewery treasures. In 1824 Peter Degraves immigrated to this southerly island where, with his skills as an architect and draftsman, he set about building an empire of small industries, the first being a sawmill in the port town of Hobart. In 1825 Degraves was imprisoned for five years on bankruptcy charges stemming from his past life in England. It was during these times Degraves designed his brewery (and also redesigned the prison).

Gold was discovered in nearby Victoria, and Degraves's various enterprises prospered. With the death of his last son, all of the Cascade enterprises became incorporated as a public company in 1883. The company bought property and expanded into the hotel and tourism industries, as well as the wine, distilled spirits, soft drink and fruit juice industries, until it ran into financial difficulties in the early 1990s. A joint venture was formed between the Cascade Group and Carlton and United Breweries. The brewery remains an important part of Tasmania's culture landscape and continues to brew Cascade Premium Lager, Cascade Special Stout, Cascade Bitter and Cascade Tiger Head.

CHARACTER DESCRIPTION OF GOLD CUP–WINNING CASCADE PALE ALE
A pale, light golden lager with a "Pale Ale" label. Clean and neutral aroma with a hint of honey character. Hop and

malt characters not detectable in aroma. Light-bodied mouth feel accompanied by a clean honeylike, light malt flavor. Refreshing with no characteristics of an ale. Hop bitterness is pleasant without being assertive. Honey character may come from controlled fermentation of a sugar adjunct. Great drinkability.

- ### Recipe for 5 U.S. *gallons* (19 *liters*) **Cascade Pale Ale**
  Targets:
  Original Gravity: 1.044 (11)
  Final Gravity: 1.006 (1.5)
  Alcohol by volume: 5.2
  Color: 3.6 SRM (7.2 EBC)
  Bittering Units: 23

ALL-GRAIN RECIPE AND PROCEDURE

5	lbs.	(2.3 kg.)	American 2-row pale malt
2	lbs.	(0.9 kg.)	light honey
¼	lb.	(114 g.)	American Cara-Pils malt
5¼	**lbs.**	**(2.4 kg.)**	**Total grains**

4   HBU   (113 MBU) Australian Pride of Ringwood hops (pellets)—60 minutes (bittering)

2   HBU   (56 MBU) American Tettnanger hops (whole)— 30 minutes (flavor)

½   oz.   (14 g.) American Tettnanger hops (whole)— steep in finished boiled wort for 2 to 3 minutes (aroma)

¼   tsp.   Irish moss

¾   c.   corn sugar for priming in bottles. Use ⅓ cup corn sugar if priming a keg.

Wyeast 2278 Czech Pils Lager yeast

A step infusion mash is employed to mash the grains. Add 5 quarts (4.8 l.) of 140-degree F (60 C) water to the crushed grain, stir, stabilize and hold the temperature at 133 degrees F (56 C) for 30 minutes. Add 2.5 quarts (2.4 l.) of boiling water, adding heat if necessary to bring temperature up to 150 degrees F (65.5 C). Hold for about 60 minutes.

After conversion, raise temperature to 167 degrees F (75 C), lauter and sparge with 2.5 gallons (9.5 l.) of 170-degree F (77 C) water. Collect about 3.5 gallons (13 l.) of runoff. Add 2 more gallons (7.6 l.) water, bittering hops and honey, and bring to a full and vigorous boil.

The total boil time will be 60 minutes. When 30 minutes remain, add flavor hops. When 10 minutes remain, add Irish moss. After a total wort boil of 60 minutes (reducing the wort volume to just over 5 gallons), turn off the heat, add aroma hops and let steep for 2 to 5 minutes. Then separate or strain out and sparge hops. Chill the wort to 65 degrees F (18 C) and direct into a sanitized fermenter. Aerate the cooled wort well. Add an active yeast culture and ferment for 4 to 6 days in the primary at 55 degrees F (15 C). Then transfer into a secondary fermenter, chill to 50 degrees F (10 C) to age for two more weeks, then lager for two to four more weeks at 40 degrees F (4.5 C).

When secondary aging is complete, prime with sugar, bottle or keg. Let condition at temperatures above 60 degrees F (15.5 C) until clear and carbonated, then store chilled.

MALT-EXTRACT RECIPE AND PROCEDURE FOR CASCADE PALE ALE

3½	lbs.	(1.6 kg.)	English light dried malt extract
2	lbs.	(0.9 kg.)	light honey
0	**lb.**	**(0 kg.)**	**Total grains**

| 5 | HBU | (142 MBU) Australian Pride of Ringwood hops (pellets)—60 minutes (bittering) |
| 2 | HBU | (56 MBU) American Tettnanger hops (whole)— 30 minutes (flavor) |

½ oz. (14 g.) American Tettnanger hops (whole)—steep in finished boiled wort for 2 to 3 minutes (aroma)

¼ tsp. Irish moss

¾ c. corn sugar for priming in bottles. Use ⅓ cup corn sugar if priming a keg.

Wyeast 2278 Czech Pils Lager yeast

Add the dried malt extract, honey and bittering hops to 2.5 gallons (9.5 l.) water and bring to a full and vigorous boil. The total boil time will be 60 minutes. When 30 minutes remain, add flavor hops. When 10 minutes remain, add Irish moss. After a total wort boil of 60 minutes, turn off the heat, add aroma hops and let steep for 2 to 5 minutes. Then separate or strain out and sparge hops, and direct the hot wort into a sanitized fermenter to which 2 gallons (7.6 l.) of cold water have been added. If necessary, add additional cold water to achieve a 5-gallon (19-l.) batch size. Chill the wort to 65 degrees F (68 C). Aerate the cooled wort well. Add an active yeast culture and ferment for 4 to 6 days in the primary at 55 degrees F (15 C). Then transfer into a secondary fermenter, chill to 50 degrees F (10 C) to age for two more weeks, then lager for two to four more weeks at 40 degrees F (4.5 C).

When secondary aging is complete, prime with sugar, bottle or keg. Let condition at temperatures above 60 degrees F (15.5 C) until clear and carbonated, then store chilled.

Silver Cup Winner

None

Bronze Cup Winner

None

Hybrid/Mixed Styles

CATEGORY 55:
AMERICAN-STYLE LAGER/
ALE OR CREAM ALE

A mild, pale, light-bodied ale, made using either a warm bottom fermentation or a top fermentation and cold, or by blending top- and bottom-fermented beers. Hop bitterness and flavor are very low. Hop aroma is often absent. Sometimes referred to as Cream Ales, these beers are crisp and refreshing. A fruity or estery aroma may be perceived. Diacetyl and chill haze should not be perceived.

Original Gravity (°Plato):   1.044–1.056 (11–14 °Plato)

Apparent Extract–
Final Gravity (°Plato):   1.004–1.010 (1–2.5 °Plato)

Alcohol by weight (volume):   3.4–4.5% (4.2–5.6%)

Bitterness (IBU):   10–22

Color (SRM):   2–5 (4–14 EBC)

Gold Cup Winner

California Blonde Ale
Coast Range Brewing Company
7050 Monterey St.
Gilroy, California, USA 95020
Brewmaster: Peter Licht
Established 1995
Production: 2,000 bbl. (2,300 hl.)

Beer and brewing are back in Gilroy, the garlic capital of the world. But there isn't any garlic in these beers. With the founding of the Coast Range Brewing Company, the Gilroy brewing tradition that began in 1868 and temporarily ended in 1919 was revived. Founded by Ron Erskine, Coast Range Brewing Company is located in a turn-of-the-century building in downtown Gilroy. It is fresh on the brewing scene but has won immediate acclaim, especially with its California Blonde Ale, which won the Gold in the World Beer Cup and received awards at the 1995 and 1996 California State Fairs. Cali-

Ron Erskine, President, and Peter Licht, Brewmaster

fornia Blonde was inspired by the New York State tradition of Genessee Cream Ale and Utica Club Cream Ale, both of which were former regional favorites of brewmaster Peter Licht. Peter was originally from Rochester, New York.

At this writing, Coast Range Brewing Company continues to expand with a bottling line. Other products produced are: Desperado Special Bitter, Blackberry Wheat Ale, India Pale Ale, Auld Lang Syne Holiday Ale, and Irish Stout.

CHARACTER DESCRIPTION OF GOLD CUP–WINNING CALIFORNIA BLONDE ALE

Crystal-clear with a deep golden color without shades of amber-orange. Beer endears with a nonassaulting yet very assertive sweet hop aroma. Not citrusy. Malt sweetness hints at a vanillalike texture, with honey character emerging afterward. A low level of DMS (a sweet corn character) softens the malt character without being perceptible on its own. A cookielike biscuit aroma evolves as the beer warms. A swirl of the beer brings out volatile sweet-floral hop aromas at any point during indulgence. First impression is full-flavored with a medium-body mouth feel, neither of which is assertive. Hop bitterness seems as though it might be of the Goldings or Fuggles type, being quite soft and earthy, but formulation indicates Perle, Cluster, Mt. Hood, and Liberty. Well balanced between malt and hop, sweet and bitter. Aftertaste has a memorable, lingering bitterness. Malt fades quickly in aftertaste.

- **Recipe for 5 U.S. gallons (19 liters) California Blonde Ale**

  Targets:
  Original Gravity: 1.046 (11.5)
  Final Gravity: 1.011 (2.7)
  Alcohol by volume: 4.6%
  Color: 6 SRM (12 EBC)
  Bittering Units: 20

ALL-GRAIN RECIPE AND PROCEDURE

8	lbs.	(3.6 kg.)	American 2-row Klages pale malt
½	lb.	(0.23 kg.)	English crystal malt—10 Lovibond
½	lb.	(0.23 kg.)	American wheat malt
¼	lb.	(114 g.)	American Victory or other aromatic malt

9¼ lbs. (4.2 kg.)   Total grains

2   HBU   (56 MBU) American Perle hops (pellets)—90 minutes (bittering)

2   HBU   (56 MBU) American Cluster hops (pellets)—90 minutes (bittering)

1   HBU   (28 MBU) American Liberty hops (pellets)—30 minutes (flavor)

2   HBU   (56 MBU) American Mt. Hood hops (pellets)—30 minutes (flavor)

½   oz.   (14 g.) American Tettnanger hops (whole)—steep in finished boiled wort for 2 to 3 minutes (aroma)

¼   tsp.   Irish moss

¾   c.   corn sugar for priming in bottles. Use ⅓ cup corn sugar if priming a keg.

Wyeast 1056 American Ale yeast

A step infusion mash is employed to mash the grains. Add 9 quarts (8.5 l.) of 140-degree F (60 C) water to the crushed grain, stir, stabilize and hold the temperature at 133 degrees F (56 C) for 30 minutes. Add 4.5 quarts (4.3 l.) of boiling water, adding heat if necessary to bring temperature up to 152 degrees F (67 C). Hold for about 60 minutes.

After conversion, raise temperature to 167 degrees F (75 C), lauter and sparge with 4 gallons (15 l.) of 170-degree F (77 C) water. Collect about 6 gallons (23 l.) of runoff, add bittering hops and bring to a full and vigorous boil.

The total boil time will be 90 minutes. When 30 minutes remain, add flavor hops. When 10 minutes remain, add Irish moss. After a total wort boil of 90 minutes (reducing the wort volume to just over 5 gallons), turn off the heat, add aroma hops and let steep for 2 to 3 minutes. Then separate or strain out and sparge hops. Chill the wort to 70 degrees F (21 C) and direct into a sanitized fermenter. Aerate the cooled wort well. Add an active yeast culture and ferment for 4 to 6 days in the primary at 65 degrees F (15 C). Then transfer into a secondary fermenter, chill to 60 degrees F (15.5 C) to age for two more weeks, then lager for two to four more weeks at 40 degrees F (4.5 C).

When secondary aging is complete, prime with sugar, bottle or keg. Let condition at temperatures above 60 degrees F (15.5 C) until clear and carbonated, then store chilled.

MASH-EXTRACT RECIPE AND PROCEDURE FOR CALIFORNIA BLONDE ALE

3¾	lbs.	(1.7 kg.)	English light dried malt extract
1¾	lbs.	(0.8 kg.)	American 2-row Klages pale malt
½	lb.	(0.23 kg.)	English crystal malt—10 Lovibond
½	lb.	(0.23 kg.)	American wheat malt
¼	lb.	(114 g.)	American Victory or other aromatic malt
3	**lbs.**	**(1.4 kg.)**	**Total grains**

2.5	HBU	(71 MBU) American Perle hops (pellets)—60 minutes (bittering)
2.5	HBU	(71 MBU) American Cluster hops (pellets)—60 minutes (bittering)
1	HBU	(28 MBU) American Liberty hops (pellets)—30 minutes (flavor)
2	HBU	(56 MBU) American Mt. Hood hops (pellets)—30 minutes (flavor)
½	oz.	(14 g.) American Tettnanger hops (whole)—steep in finished boiled wort for 2 to 3 minutes (aroma)
¼	tsp.	Irish moss

¾ c.   corn sugar for priming in bottles. Use ⅓ cup
         corn sugar if priming a keg.
Wyeast 1056 American Ale yeast

A step infusion mash is employed to mash the grains. Add 3 quarts (2.9 l.) of 140-degree F (60 C) water to the crushed grain, stir, stabilize and hold the temperature at 133 degrees F (56 C) for 30 minutes. Add 1.5 quarts (1.5 l.) of boiling water, add heat to bring temperature up to 152 degrees F (67 C) and hold for about 60 minutes.

After conversion, raise temperature to 167 degrees F (75 C), lauter and sparge with 1.5 gallons (5.7 l.) of 170-degree F (77 C) water. Collect about 2.5 gallons (9.5 l.) of runoff. Add malt extract and bittering hops and bring to a full and vigorous boil.

The total boil time will be 60 minutes. When 30 minutes remain, add flavor hops. When 10 minutes remain, add Irish moss. After a total wort boil of 60 minutes, turn off the heat, add aroma hops and let steep 2 to 5 minutes. Then separate or strain out and sparge hops, and direct the hot wort into a sanitized fermenter to which 2 gallons (7.6 l.) of cold water have been added. If necessary, add additional cold water to achieve a 5-gallon (19-l.) batch size. Chill the wort to 70 degrees F (21 C). Aerate the cooled wort well. Add an active yeast culture and ferment for 4 to 6 days in the primary at 65 degrees F (15 C). Then transfer into a secondary fermenter, chill to 60 degrees F (15.5 C) to age for two more weeks, then lager for two to four more weeks at 40 degrees F (4.5 C).

When secondary aging is complete, prime with sugar, bottle or keg. Let condition at temperatures above 60 degrees F (15.5 C) until clear and carbonated, then store chilled.

Silver Cup Winner

None

Bronze Cup Winner

Point Pale Ale
Stevens Point Brewery
2617 Water St.
Stevens Point, Wisconsin, USA 54481

Orange-hued amber color. A complex aroma emerges with a textured fruitiness that bewilders with a combination of elegantly floral hops, malt and caramel sweetness. Medium- to light-bodied mouth feel and ale fruitiness are accompanied by a pleasantly clean yet not too sweet malt character. A bit of caramel emerges between thoughts. Willamette hop character seems to dominate over Cascade. Soft hop flavor and gentle bitterness provide a clean, inviting and excellent balance of characters.

Brewery formulation uses Cascade and Willamette hops according to the label.

Estimated profile based on tasting
Color: 7–9 SRM (14–18 EBC)
Bittering Units: 22–26

CATEGORY 56:
AMERICAN-STYLE WHEAT ALE OR LAGER

This beer can be made using either an ale or lager yeast. Brewed with 30 to 50 percent wheat, hop rates are higher and carbonation is lower than with German-Style Wheat Beers. A fruity-estery aroma and flavor are typical but at low levels; however, phenolic, clovelike characteristics should not be perceived. Color is usually golden to light amber, and the body light to medium in character. Diacetyl should be at very low levels.

Original Gravity (°Plato):	1.030–1.050 (9.5–12.5 °Plato)
Apparent Extract– Final Gravity (°Plato):	1.004–1.018 (1–4.5 °Plato)
Alcohol by weight (volume):	2.8–3.6% (3.5–4.5%)
Bitterness (IBU):	5–17
Color (SRM):	2–8 (4–16 EBC)

Gold Cup Winner

Thomas Kemper Hefeweizen
Thomas Kemper Brewing Company
91 S. Royal Brougham Way
Seattle, Washington, USA 98134
Brewmaster: Rande Reed
Established 1985
Production: 123,000 bbl. (144,000 hl.) total for Pyramid
Breweries Inc., parent company of Thomas Kemper

"I live for beer."
Rande Reed, brewmaster at
the Thomas Kemper Brewing
Company, speaks and enacts
these words without jest. If
ever a brewmaster made a
difference, Rande is one. The
Thomas Kemper Brewing
Company was founded in
1985, while Rande Reed was
still homebrewing in Milwaukee, Wisconsin. His love of
beer took him on travels to
Europe, learning the secrets
and arts of a long tradition.

Brian J. Marin

Rande Reede, Brewmaster

Veteran homebrewers will recall Rande's expert articles on home-brewed cask-conditioned real ale. His were the first to appear on the subject in the American Homebrewers Association's *Zymurgy* magazine back in the early 1980s. Progressing from brewing ale to brewing lager, Rande left his job as a steelworker in 1988 for half the pay at the Sprecher Brewing Company in Milwaukee. Soon thereafter he found himself moving to Poulsbo, Washington, a tiny community on Washington State's Olympic peninsula and the home of the tiny Thomas Kemper Brewing Company.

The lager brewery founded by Will Kemper and Andrew Thomas was unique in the ale-loving Northwest. Now the brewery enjoys a partnership with parent company Hart Brewing (brewers of Pyramid beers). Sales have increased dramatically over the past few years, the brewery has been upgraded with modern equipment, and Rande Reed continues to meet the challenge of the Northwest, where Thomas Kemper still remains one of the very few lager microbreweries in the area.

Thomas Kemper Brewing Company brews seven different lagers and other specialties including White Beer, Wiezen-Berry, Dark Lager, Helles Blueberry Lager, Winterbrau, Mai-Bock, Bohemian Dunkel, Honey Weizen, Pale Lager, Roggen Rye, Amber Lager and Oktoberfest.

CHARACTER DESCRIPTION OF GOLD CUP–WINNING THOMAS KEMPER HEFEWEIZEN

Light golden yellow color. Light, soft, biscuitlike wheat-malt aroma with a sweet floral hop character. Though bottle-conditioned (*Hefe-weizen* translated from the German means "with yeast"), no degree of yeast character is evident in aroma. While not assertive, hops maintain a primary aromatic role. First encounter impresses with a clean, neutral, well-balanced flavor profile of soft hops and soft malt. Memorable impression of soft hop bitterness, flavor and aroma are established. Medium-bodied mouth feel complements the balance of ingredients.

• **Recipe for 5 U.S. gallons (19 liters) Thomas Kemper Hefeweizen**

Targets:
Original Gravity: 1.060 (15)
Final Gravity: 1.014 (3.5)
Alcohol by volume: 6%
Color: 14 SRM (28 EBC)
Bittering Units: 60

ALL-GRAIN RECIPE AND PROCEDURE

3½	lbs. (1.6 kg.)	American 2-row Klages pale malt
5	lbs. (2.3 kg.)	American wheat malt
8½	**lbs. (3.9 kg.)**	**Total grains**

2.5	HBU	(71 MBU) American Nugget hops (pellets)—60 minutes (bittering)
4	HBU	(113 MBU) American Tettnanger hops (pellets)—30 minutes (flavor)
½	oz.	(14 g.) American Liberty hops (pellets)—steep in finished boiled wort for 2 to 3 minutes (aroma)

¼	tsp.	Irish moss
¾	c.	corn sugar for priming in bottles. Use ⅓ cup corn sugar if priming a keg.

Wyeast 1335 British Ale yeast II; dry, crisp, with malty finish, good flocculation, low esters.

A single-step infusion mash is employed to mash the grains. Add 8.5 quarts (8.1 l.) of 170-degree F (77 C) water to the crushed grain, stir, stabilize and hold the temperature at 153 degrees F (67 C) for 60 minutes.

After conversion, raise temperature to 167 degrees F (75 C), lauter and sparge with 4 gallons (15 l.) of 170-degree F (77 C) water. Collect about 5.5 gallons (21 l.) of runoff, add bittering hops and bring to a full and vigorous boil.

The total boil time will be 60 minutes. When 30 minutes

remain, add flavor hops and Irish moss. After a total wort boil of 60 minutes (reducing the wort volume to just over 5 gallons), turn off the heat, add aroma hops and let steep 2 to 5 minutes. Then separate or strain out and sparge hops. Chill the wort to 70 degrees F (21 C) and direct into a sanitized fermenter. Aerate the cooled wort well. Add an active yeast culture and ferment for 4 to 6 days in the primary, maintaining a temperature of 70 degrees F (21 C). Then transfer into a secondary fermenter, chill to 65 degrees F (18 C) and age for two to three weeks.

When secondary aging is complete, prime with sugar, bottle or keg. Let condition at temperatures above 60 degrees F (15.5 C) until clear and carbonated.

MASH-EXTRACT RECIPE AND PROCEDURE FOR THOMAS KEMPER HEFEWEIZEN

3¼ lbs.	(1.5 kg.)	wheat malt extract syrup (50% barley, 50% wheat)
1¾ lbs.	(0.8 kg.)	American 2-row Klages pale malt
2½ lbs.	(1.1 kg.)	American wheat malt
4¼ lbs.	**(1.9 kg.)**	**Total grains**

3	HBU	(85 MBU) American Nugget hops (pellets)—60 minutes (bittering)
4	HBU	(113 MBU) American Tettnanger hops (pellets)—30 minutes (flavor)
½	oz.	(14 g.) American Liberty hops (pellets)—steep in finished boiled wort for 2 to 3 minutes (aroma)

¼	tsp.	Irish Moss
¾	c.	corn sugar for priming in bottles. Use ⅓ cup corn sugar if priming a keg.

Wyeast 1335 British Ale yeast II; dry, crisp, with malty finish, good flocculation, low esters.

A single-step infusion mash is employed to mash the grains. Add 4.25 quarts (4 l.) of 170-degree F (77 C) water to the crushed grain, stir, stabilize and hold the temperature at 153 degrees F (67 C) for 60 minutes.

After conversion, raise temperature to 167 degrees F (75 C), lauter and sparge with 2.5 gallons (9.5 l.) of 170-degree F (77 C) water. Collect about 2.5 gallons (9.5 l.) of runoff. Add malt extract and bittering hops and bring to a full and vigorous boil.

The total boil time will be 60 minutes. When 30 minutes remain, add flavor hops. When 10 minutes remain, add Irish moss. After a total wort boil of 60 minutes, turn off the heat, add aroma hops and let steep 2 to 5 minutes. Then separate or strain out and sparge hops, and direct the hot wort into a sanitized fermenter to which 2 gallons (7.6 l.) of cold water have been added. If necessary, add additional cold water to achieve a 5-gallon (19-l.) batch size. Chill the wort to 70 degrees F (21 C). Aerate the cooled wort well. Add an active yeast culture and ferment for 4 to 6 days in the primary, maintaining temperature of 70 degrees F (21 C). Then transfer into a secondary fermenter, chill to 65 degrees F (18 C) and age for two to three weeks.

When secondary aging is complete, prime with sugar, bottle or keg. Let condition at temperatures above 60 degrees F (15.5 C) until clear and carbonated.

Silver Cup Winner

Weiss Guy Wheat
Alcatraz Brewing Co.
Indianapolis, Indiana, USA

Light amber-orange color with yeast and chill haze. A slight, almost resinlike, citrusy hop aroma is evident, but not overdone. Mouth feel is medium-bodied with the somewhat harsh hop flavor of a higher alpha-acid hop. Some fruitiness from hops and fermentation, but retreats to a neutral flavor impression. Sweet malt character with the possibility of Vienna, Munich or caramel malts is secondary to fruitiness, followed by a bitter aftertaste. Persistent aroma of hops carries through the tasting experience. Generally a neutral beer with malt and hops lending the primary character. Fermentation characteristics are minimal. Pleasantly clean finish with a notable lingering bitterness.

Brewery formulation uses English light Carastan and Munich and American wheat malts, with American Perle, Tettnanger and Ultra hops for bitterness, flavor and aroma.

Estimated profile based on tasting
Original Gravity: 1.053 (13.3) indicated by the brewery
Final Gravity: 1.011 (2.8) indicated by the brewery
Alcohol by volume: 6.4% indicated by the brewery
Color: 5–7 SRM (10–14 EBC) (5 SRM indicated by the brewery)
Bittering Units: 25–30 (18 indicated by the brewery)

Bronze Cup Winner

Red Ass Honey Wheat
Red Ass Brewing Co.
Ft. Collins, Colorado, USA

Light golden amber color. Aroma presents a full dose of floral American hops and pale malt sweetness. First flavor impressions portray a full, sweet character and medium-bodied mouth feel. There may be a slight hint of caramel in the flavor. Soft hop bitterness emerges, balancing the initial bigness of malt. Gentle bitterness lingers in the aftertaste after the impact of malt and honey fade. Smooth, well balanced, and once again, as with almost all of the winning beers, this beer is clean. The soft bitterness grows the more one indulges. There may be some hop flavor, but its contribution is lost in the complexity of overall smoothness and balance, hence not emerging on its own.

Estimated profile based on tasting
Color: 5–7 SRM (10–14 EBC)
Bittering Units: 29–33

CATEGORY 57: FRUIT BEERS

Fruit Beers are any beers using fruit as an adjunct in either primary or secondary fermentation, providing obvious yet harmonious fruit qualities. Fruit qualities should not be overpowered by hop character. If a fruit (such as juniper berry) has an herbal or spice quality, it should be entered into the Herb and Spice Beers category.

Original Gravity (°Plato):	1.030–1.110 (7.5–27.5 °Plato)
Apparent Extract– Final Gravity (°Plato):	1.006–1.030 (1.5–7.5 °Plato)
Alcohol by weight (volume):	2–9.5% (2.5–12%)
Bitterness (IBU):	5–70
Color (SRM):	5–50 (10–200 EBC)

Gold Cup Winner

Liefmans Frambozen
Brouwerij Liefmans (Liefmans Brewery)
Aalstrstraat 200
9700 Oudenaarde, Belgium
Brewmaster: Filip de Velder
Established 1679
Production: 25,600 bbl. (30,000 hl.)

Annick De Splenter

The wrapping of the bottles

Liefmans also won the Belgian-Style Flanders/Oud Bruin Ales category. Please see that category for a description of the brewery.

CHARACTER DESCRIPTION OF GOLD CUP–WINNING LIEFMANS
FRAMBOZEN

Perhaps a December night by a stone fireplace in Belgium. A sensual brown ale with sparkling ruby red hues. A corked beer with intensely uncorked raspberry aromatics. Aroma can be described as a complex combination of raspberry, cassis (black currant), and balsam (fir), with a cork-generated earthy mustiness inspiring endless inhalation. Complexity of characters due to the fermentation and aging process evolves to this lascivious indulgence. Frambozen becomes an emotional experience, if one lets it. Fruitiness is a low tone, combining with the skillfully fashioned, subdued maltiness of a brown ale. This raspberry ale is almost so complex and sophisticated that only the original burst of aroma suggests raspberry. For those who appreciate the complexities of red fruit, this blend is suggestive of cherries and blue cassis contending with raspberries. All this and I haven't even tasted it yet! The smell is easily worth the experience alone.

Flavor is soft and not excessively acidic. Malts create a medium-bodied mouth feel. Softness of caramel underpins the higher fruity flavor notes. Clean aftertaste with a suggestion of dryness in the finish. Hop bitterness is very low.

- ### Recipe for 5 U.S. gallons (19 liters) Liefmans Frambozen
  Targets:
  Original Gravity: 1.053 (13.3)
  Final Gravity: 1.020 (5)
  Alcohol by volume: 4.4%
  Color: 18 SRM (36 EBC)
  Bittering Units: 22

ALL-GRAIN RECIPE AND PROCEDURE

3¾	lbs.	(1.7 kg.)	French 2-row Prisma or Belgian pale ale malt
3¾	lbs.	(1.7 kg.)	French or German Munich malt—7 Lovibond (15 EBC)
1	lb.	(0.45 kg.)	French or Belgian CaraMunich malt—70 Lovibond (150 EBC)
1	lb.	(0.45 kg.)	rice
2½	oz.	(70 g.)	Belgian Special "B" crystal malt
3	lbs.	(1.4 kg.)	ripe raspberries (or natural raspberry extract measured to taste)
½	lb.	(0.23 kg)	black currants (cassis)
9.7	**lbs.**	**(4.4 kg.)**	**Total grains**

4	HBU	(113 MBU)	English Challenger hops (pellets)—90 minutes (bittering)
1	HBU	(28 MBU)	Polish Lublin or Styrian Goldings

hops (pellets)—10 minutes (aroma)

¼   tsp.   Irish moss
¾   c.   corn sugar for priming in bottles. Use ⅓ cup
              corn sugar if priming a keg.
Recommend beginning fermentation with a fruity ale yeast
such as Wyeast 1098 and then introducing cultured yeast from
a bottle of Liefmans Goudenband or other live Flanders brown
ale when beer has rested in the secondary.

A rice-cooking regime along with a step infusion mash is
employed to mash the grains. Crush and mill rice into small
pieces. Add crushed rice to 1 gallon (3.8 l.) of water and
boil for 20 minutes. Add 1.5 more gallons (5.7 l.) of water to
cooked rice mash to achieve a temperature of 130 degrees F
(54.5 C). Add malt and hold the temperature at 122 degrees
F (50 C) for 30 minutes. Add 5 quarts (4.8 l.) of boiling
water, adding heat if necessary to bring temperature up to
150 degrees F (65.5 C). Hold for about 60 minutes.

After conversion, raise temperature to 167 degrees F (75
C), lauter and sparge with 4 gallons (15 l.) of 170-degree F
(77 C) water. Collect about 6 gallons (23 l.) of runoff, add
bittering hops and bring to a full and vigorous boil.

The total boil time will be 90 minutes. When 10 minutes
remain, add aroma hops and Irish moss. After a total wort boil
of 90 minutes (reducing the wort volume to just over 5 gallons), turn off the heat, then separate or strain out and sparge
hops. Chill the wort to 70 degrees F (21 C) and direct into a
sanitized fermenter. Aerate the cooled wort well. Add an active
yeast culture and ferment for 4 to 6 days in the primary at 70
degrees F (21 C). Then transfer into a secondary fermenter and
add crushed raspberries, black currants, and cultured "Goudenband" yeast if available. Ferment for an additional two to
three weeks. Then transfer to a third vessel and let age for two
to three more weeks at 70 F (21 C). If raspberry extract is used,
then aging in a third vessel is not necessary.

When aging is complete, prime with sugar, bottle or keg.

Closing with a cork will enhance the character of this beer. Wire the cork securely to the bottle. Let condition at temperatures above 65 degrees F (18 C) until clear and carbonated, then store chilled.

MASH-EXTRACT RECIPE AND PROCEDURE FOR LIEFMANS FRAMBOZEN

2	lbs.	(0.9 kg.)	English amber dried malt extract
1½	lbs.	(0.7 kg.)	French 2-row Prisma or Belgian pale ale malt
1½	lbs.	(0.7 kg.)	French or German Munich malt—7 Lovibond (15 EBC)
1	lb.	(0.45 kg.)	French or Belgian CaraMunich malt—70 Lovibond (150 EBC)
1	lb.	(0.45 kg.)	rice
2½	oz.	(70 g.)	Belgian Special "B" crystal malt
3	lbs.	(1.4 kg.)	ripe raspberries (or natural raspberry extract measured to taste)
½	lb.	(0.23 kg.)	black currants (cassis)
5.2	**lbs.**	**(2.4 kg.)**	**Total grains**

5 HBU (142 MBU) English Challenger hops (pellets)—75 minutes (bittering)

1 HBU (28 MBU) Polish Lublin or Styrian Goldings hops (pellets)—10 minutes (aroma)

¼ tsp. Irish moss

¾ c. corn sugar for priming in bottles. Use ⅓ cup corn sugar if priming a keg.

Recommend beginning fermentation with a fruity ale yeast such as Wyeast 1098 and then introducing cultured yeast from a bottle of Liefmans Goudenband or other live Flanders brown ale when beer has rested in the secondary.

A rice-cooking regime along with a step infusion mash is employed to mash the grains. Crush and mill rice into small pieces. Add crushed rice to 2 quarts (1.8 l.) of water and boil for 20 minutes. Add 3 more quarts (2.9 l.) of water to cooked rice mash to achieve a temperature of 130 degrees F (54.5 C). Add malt and hold the temperature at 122 degrees F (50 C) for 30 minutes. Add 2.5 quarts (2.4 l.) of boiling water, adding heat if necessary to bring temperature up to 150 degrees F (65.5 C). Hold for about 60 minutes.

After conversion, raise temperature to 167 degrees F (75 C), lauter and sparge with 2.5 gallons (9.5 l.) of 170-degree F (77 C) water. Collect about 3.5 gallons (13 l.) of runoff, add bittering hops and bring to a full and vigorous boil.

The total boil time will be 75 minutes. When 10 minutes remain, add aroma hops and Irish moss. After a total wort boil of 75 minutes (reducing the wort to 2.5 to 3 gallons [9.5–11.4 l.], turn off the heat, then separate or strain out and sparge hops, and direct the hot wort into a sanitized fermenter to which 2 gallons (7.6 l.) of cold water have been added. If necessary, add additional cold water to achieve a 5-gallon (19-l.) batch size. Chill the wort to 70 degrees F (21 C). Aerate the cooled wort well. Add an active yeast culture and ferment for 4 to 6 days in the primary at 70 degrees F (21 C). Then transfer into a secondary fermenter and add crushed raspberries, black currants and cultured "Goudenband" yeast if available. Ferment for an additional two to three weeks. Then transfer to a third vessel and let age for two to three more weeks at 70 F (21 C). If raspberry extract is used, then aging in a third vessel is not necessary.

When aging is complete, prime with sugar, bottle or keg. Closing with a cork will enhance the character of this beer. Wire the cork securely to the bottle. Let condition at temperatures above 65 degrees F (18 C) until clear and carbonated, then store chilled.

Silver Cup Winner

Belgian Red Wisconsin Cherry
New Glarus Brewing Co.
New Glarus, Wisconsin, USA

Bright red color glamorizes this award-winning beer. Intensity of the cherry aroma is a giant component. Cherry character is reminiscent of wild cherries and ripe, flavorful pie cherries. No earthiness or mustiness as is usually associated with corked or Belgian-made cherry beers. This cherry beer stands alone on the quality of the cherries, which are unique in their intensity and fresh, lively, and spicy in aroma. Medium-bodied mouth feel. Flavor impact is entirely the sweetness and sourness of cherries. Hop bitterness is negligible. The very round character of the sweetness dismisses any notion of artificial ingredients or process. Malt flavor is not obvious, but is evident in establishing a foundation and texture with a bit of roast character that does not emerge specifically. Without the malt foundation, I fear the sourness and sweetness might otherwise escape to the outward reaches of the solar system. Clean and simple yet brilliantly satisfying and complex. A dry, clean finish despite the medium body texture.

According to the label, brewery formulation uses one pound of Door County Montmorency cherries in every 750-milliliter bottle, wheat and roasted Belgian malts, and hops aged one year in the brewery.

Estimated profile based on tasting
Color: Red SRM (Red EBC)
Bittering Units: 12–13

Bronze Cup Winner

Liefmans Kriek
Brouwerij Liefmans
Oudenaarde, Belgium

A medium brown color whose primary character is a tawny amber-red. A Belgian corked ale flavored with cherries, Liefmans Kriek presents an incredibly erotic aroma reminiscent of things old, musty, and antique. Wild cherry aroma married with a gentle golden earthiness recalls pleasant memories. In the ancient tradition of making this type of beer, cherries are a huge part of the aroma and flavor. Medium-bodied mouth feel complemented by an assertive acidity promoting the fruitiness of the cherries. No hop bitterness whatsoever in flavor or aftertaste. Tart, clean, dry finish with a continuing acidic aftertaste that trails off to a memory of cherries and earthiness.

Brewery formulation uses French Munich, CaraMunich, and CaraVienne malts, rice and cherries, with English Challenger and Goldings hops for bitterness and aroma.

Estimated profile based on tasting
Original Gravity: 1.058 (14.4) indicated by the brewery
Final Gravity: 1.018 (4.5) indicated by the brewery
Alcohol by volume: 6.5%
Color: 18–21 SRM (36–42 EBC) (75 EBC indicated by the brewery)
Bittering Units: 18 indicated by the brewery

CATEGORY 58:
HERB AND SPICE BEERS

Herb Beers use herbs or spices (derived from roots, seeds, fruits, vegetables, flowers, etc.) other than hops to create a distinct character. Underhopping allows the spice or herb to contribute to the flavor profile.

Original Gravity (°Plato):	1.030–1.110 (7.5–27.5 °Plato)
Apparent Extract–	
Final Gravity (°Plato):	1.006–1.030 (1.5–7.5 °Plato)
Alcohol by weight (volume):	2–9.5% (2.5–12%)
Bitterness (IBU):	5–70
Color (SRM):	5–50 (10–200 EBC)

Gold Cup Winner

Coriander Rye Ale
Bison Brewing Company
2598 Telegraph Ave.
Berkeley, California, USA 94704
Brewmaster: Scott Meyer
Established 1988
Production: 1,000 bbl. (1,200 hl.)

Berkeley, California. The very ground from which many fresh and eclectic ideas have originated helps inspire the spirit of this small brewpub tucked away along Telegraph Avenue. "Alternative nature," the brewery claims of its surroundings. A truly homebrewed spirit permeates its list of "generally" available products. I emphasize "generally" because at any given moment you'll find a new experimental brew on tap. Once a month the brewery bottles one of its beers and offers it in limited quantities, sold in the pub and area outlets. All of its beers are presently live and unfiltered— and the special qualities that bottle conditioning contributes are evident in the World Cup–winning Coriander Rye Ale and Extra Special Bitter.

The list of other products speaks for the spirit of this quality-minded brewery. It includes India Pale Ale, Choco-

Scott Meyer

late Stout, Alder Smoked Scotch Ale, Extra Special Bitter, Coriander Rye Ale, Pumpernickel Ale, Juniper Smoked Ale, Honey Basil Ale, Lemongrass Wheat Ale, Toasted Oat Molasses Brown Ale, Gingerbread Ale and several others worth exploring.

From the label:

Hippocrates' forehead furled for a moment as he carefully examined his patient. Slowly at first, then more deliberately, he began to write a prescription on a scrap of goat parchment.

The patient, perplexed, squinted at the parchment, then a smile lit up his face as he realized the prescription was a recipe for Coriander Rye Ale. He knew just what to do, and he assembled his brewing amphorae and built his fire . . .

Later, when the salubrious effects were known to all who partook of this fine, fragrant, copper-hued ale, the patient was feeling much better.

Later, much later, nearly two and a half millenniums or so, while the brewers at Bison Brewing Company were attempting to read their own scribbled notes, they went ahead and brewed another batch of Coriander Rye Ale . . .

Just what the doctor ordered!

And now you've heard . . . the other side of the story.

CHARACTER DESCRIPTION OF GOLD CUP–WINNING CORIANDER RYE ALE

Medium amber, well-carbonated, bottle-conditioned beer. Great head retention from which emerges a fruity, floral, sweet, spicy, Centennial/Cascade-type hop aroma. Extremely complex and fun to smell. Coriander is very subtle and blends in with the hop aroma to express something quite apart from each individual character.

POW! The coriander flavor comes at you but does not attack. Mouth feel/body is low-medium and finishes quite dry. Slight astringency, possibly from rye? Bitterness is soft medium to low in intensity, not expressing an assertive character, yet its profile is evident in the finish and after-taste. As the beer warms, a toasty biscuitlike malt character emerges in aroma and flavor. Hop flavor is one consistent theme throughout the indulgence. Generally sweet finish with a brief finale that includes a bite of rye, coriander and hop bitterness.

- **Recipe for 5 U.S. gallons (19 liters) Coriander Rye Ale**
  Targets:
  Original Gravity: 1.052 (13)
  Final Gravity: 1.014 (3.5)
  Alcohol by volume: 5.2%
  Color: 9 SRM (18 EBC)
  Bittering Units: 28

ALL-GRAIN RECIPE AND PROCEDURE

6¼ lbs.	(2.8 kg.)	Canadian 2-row Harrington pale malt
1½ lbs.	(0.7 kg.)	rye malt
1 lb.	(0.45 kg.)	Canadian Munich malt
¼ lb.	(114 g.)	Belgian Carastan malt—34 Lovibond
½ lb.	(0.23 kg.)	Belgian Carastan malt—15 Lovibond
¼ lb.	(114 g.)	American Victory or other aromatic malt
9¾ lbs.	**(4.4 kg.)**	**Total grains**

2.5 HBU (71 MBU) American Centennial hops (whole)—90 minutes (bittering)

2	HBU	(56 MBU) German Northern Brewer hops (whole)—90 minutes (bittering)
4	HBU	(113 MBU) German Northern Brewer hops (whole)—30 minutes (flavor)
¼	oz.	(7 g.) American Centennial hops (pellets)—steep in finished boiled wort for 2 to 3 minutes (aroma)
¼	oz.	(7 g.) freshly rushed coriander seed
¼	tsp.	Irish moss
¾	c.	corn sugar for priming in bottles. Use ⅓ cup corn sugar if priming a keg.

Wyeast 1098 Ale yeast (Whitbread origin)

A single-step infusion mash is employed to mash the grains. Add 10 quarts (9.5 l.) of 170-degree F (77 C) water to the crushed grain, stir, stabilize and hold the temperature at 153 degrees F (67 C) for 60 minutes.

After conversion, raise temperature to 167 degrees F (75 C), lauter and sparge with 4.5 gallons (17 l.) of 170-degree F (77 C) water. Collect about 6 gallons (23 l.) of runoff, add bittering hops and bring to a full and vigorous boil.

The total boil time will be 90 minutes. When 30 minutes remain, add flavor hops. When 10 minutes remain, add crushed coriander and Irish moss. After a total wort boil of 90 minutes (reducing the wort volume to just over 5 gallons), turn off the heat, add aroma hops and let steep 2 to 5 minutes. Then separate or strain out and sparge hops. Chill the wort to 65 to 70 degrees F (18–21 C) and direct into a sanitized fermenter. Aerate the cooled wort well. Add an active yeast culture and ferment for 4 to 6 days in the primary. Then transfer into a secondary fermenter and let age for three more weeks at temperatures of 65 to 70 degrees F (68.5–21 C)

When secondary aging is complete, prime with sugar, bottle or keg. Let condition at temperatures above 60 degrees F (15.5 C) until clear and carbonated.

MASH-EXTRACT RECIPE AND PROCEDURE FOR CORIANDER RYE ALE

3	lbs.	(1.4 kg.)	English light dried malt extract
2	lbs.	(0.9 kg.)	Canadian 2-row Harrington pale malt
1	lb.	(0.45 kg.)	rye malt
1	lb.	(0.45 kg.)	Canadian Munich malt
¼	lb.	(114 g.)	Belgian Carastan malt—34 Lovibond
½	lb.	(0.23 kg.)	Belgian Carastan malt—15 Lovibond
¼	lb.	(114 g.)	American Victory or other aromatic malt
5	**lbs.**	**(2.3 kg.)**	**Total grains**

2.5 HBU (71 MBU) American Centennial hops (whole)—75 minutes (bittering)

3 HBU (85 MBU) German Northern Brewer hops (whole)—75 minutes (bittering)

4 HBU (113 MBU) German Northern Brewer hops (whole) —30 minutes (flavor)

¼ oz. (7 g.) American Centennial hops (pellets)— steep in finished boiled wort for 2 to 3 minutes (aroma)

¼ oz. (7 g.) freshly crushed coriander seed

¼ tsp. Irish moss

¾ c. corn sugar for priming in bottles. Use ⅓ cup corn sugar if priming a keg.

Wyeast 1098 Ale yeast (Whitbread origin)

A single-step infusion mash is employed to mash the grains. Add 5 quarts (4.8 l.) of 170-degree F (77 C) water to the crushed grain, stir, stabilize and hold the temperature at 153 degrees F (67 C) for 60 minutes.

After conversion, raise temperature to 167 degrees F (75 C), lauter and sparge with 2.5 gallons (9.5 l.) of 170-degree F (77 C) water. Collect about 3 gallons (11.5 l.) of runoff, add bittering hops and bring to a full and vigorous boil.

The total boil time will be 75 minutes. When 30 minutes remain, add flavor hops. When 10 minutes remain, add crushed coriander and Irish moss. After a total wort boil of 75 minutes, turn off the heat, add aroma hops and let steep 2 to 5 minutes. Then separate or strain out and sparge hops, and direct the hot wort into a sanitized fermenter to which 2 gallons (7.6 l.) of cold water have been added. If necessary, add additional cold water to achieve a 5-gallon (19-l.) batch size. Chill the wort to 65 to 70 degrees F (68.5–21 C). Aerate the cooled wort well. Add an active yeast culture and ferment for 4 to 6 days in the primary. Then transfer into a secondary fermenter and leg age for three more weeks at temperatures of 65 to 70 degrees F (68.5–21 C).

When secondary aging is complete, prime with sugar, bottle or keg. Let condition at temperatures above 60 degrees F (15.5 C) until clear and carbonated.

Silver Cup Winner

RSB Spiced Scotch Ale
Routh Street Brewery
Dallas, Texas, USA

Deep brown color with sparkling red hues. Exotic aroma is reminiscent of anise, wintergreen and a hint of licorice. (See below for the real things.) Hops also come through gently with a floral nature. Very full bodied mouth feel accompanied by full-flavored malt sweetness. Hops and herb flavors come on as a secondary element. A bit of dry spiciness in flavor from roasted malts, giving a hint of astringency to an otherwise very full bodied beer. It is the malt that really impacts the palate and plays the major role in the aftertaste. Though the beer is not bitter, the hops certainly serve to balance the full malt texture and flavor. And while there is some bitterness in the flavor and aftertaste that emerge after the initial assault of malt, it does not overtake the intriguing spice/anice/floral character and full malt balance. Some darker roasted malts may also be involved, revealed by a hint of their astringency and bite. Overall a very intriguing, memorable beer.

Brewery formulation uses American roast barley, Belgian Biscuit malts, orange peel, coriander, ginger and allspice, with English Kent Goldings and Hallertauer Northern Brewer hops for bitterness and flavor. (Spices fooled me.)

Estimated profile based on tasting
Original Gravity: 1.061 (15.2) indicated by the brewery
Final Gravity: 1.012 (3.1) indicated by the brewery
Alcohol by volume: 6.7% indicated by the brewery
Color: 18–21 SRM (36–42 EBC) (18 indicated by the brewery)
Bittering Units: 33–37 (26 indicated by the brewery)

Bronze Cup Winner

Wit Amber
Spring Street Brewing Co.
New York, New York, USA

Medium amber color with an orange hue. Full, sweet caramel/crystal malt aroma pleasantly combines with floral character of coriander and orange peel. But besides the traditionally Belgian aromatic characters, there are others that suggest gentle doses of nutmeg and cinnamon . . . and then a bit more, dare I guess, almost a subtle lavenderlike character. Medium-bodied mouth feel with a refreshing finish. Caramel malts serve to blend, soften and combine all of the gentle herbal qualities, resulting in a very complex and drinkable brew. Hop bitterness is gentle, soft and easy on the palate. The caramel malts and spices are the main aspects of the overall character. The bitterness is a fulcrum between spice, body and malt. A beer that deserves attention.

Estimated profile based on tasting
Color: 7–9 SRM (14–18 EBC)
Bittering Units: 20–24

CATEGORY 59: SPECIALTY BEERS

These beers are brewed using unusual fermentables other than, or in addition to, malted barley. For example, maple syrup, potatoes or honey would be considered unusual. Rice, corn or wheat are not considered unusual.

Original Gravity (°Plato):   1.030–1.110 (7.5–27.5 °Plato)

Apparent Extract–
Final Gravity (°Plato):   1.006–1.030 (1.5–7.5 °Plato)

Alcohol by weight (volume):   2–9.5% (2.5–12%)

Bitterness (IBU):   0–100

Color (SRM):   1–100 (3–400 EBC)

Gold Cup Winner

B&H Breakfast Toasted Ale
Barley & Hopps
201 South B Street
San Mateo, California, USA 94401
Brewmaster: R. J. Trent
Established 1995
Production: 1,400 bbl. (1,600 hl.)

The brewery's grandeur is enveloped in a 16,500-square-foot, two-story restored Art Deco building in downtown San Mateo. Barley & Hopps is more than just a brewery, featuring a smokehouse, an extensive game room, banquet facilities, a smoking lounge that offers single-malt scotch, fine ports and cigars, and a live blues club.

Brewmaster R. J. Trent is a certified beer judge and veteran homebrewer who turned pro to make the popular beers at Barley & Hopps. World Beer Cup winner B&H Breakfast Toasted Ale was created with the unique addition of oats

R. J. Trent

toasted in the brewery's pizza oven, giving the ale "a sweet, smoky, chocolate flavor."

Other brews found at Barley & Hopps brewpub include Golden Pale Ale, Season Fruit Wheat, Brewmaster's Special, Oatmeal Stout, Rye Pale Ale, Roaring Red Ale and Pale Oatmeal Honey Wheat. A limited selection is located on draft at other nearby locations.

CHARACTER DESCRIPTION OF GOLD CUP–WINNING B&H BREAKFAST TOASTED ALE

An amber beer with a red hue. Crystal-clear. Toasted-malt aroma and subtle sweet, floral hop character. Flavor has a slightly peppery, astringent character with a hint of diacetyl. Medium-bodied ale with a fair degree of attenuated dry finish. Aromatic, caramel, toasted and wheat malt characters are evident. Caramel malt follows in aftertaste with some astringent bitterness. Hop bitterness contributes an assertiveness in overall balance as well as in aftertaste.

- *Recipe for 5 U.S. gallons (19 liters)* B&H *Breakfast Toasted Ale*

Targets:

Original Gravity: 1.052 (13)

Final Gravity: 1.016 (4)

Alcohol by volume: 5%

Color: 8 SRM (16 EBC)

Bittering Units: 28

ALL-GRAIN RECIPE AND PROCEDURE

6¼	lbs.	(2.8 kg.)	American 2-row Klages pale malt
1¼	lbs.	(0.6 kg.)	American wheat malt
1½	lbs.	(0.7 kg.)	American Victory malt
1	lb.	(0.45 kg.)	oven-toasted flaked oats
½	lb.	(0.23 kg.)	English crystal malt—40 Lovibond
¼	lb.	(114 g.)	Belgian Biscuit malt
10¾	**lbs.**	**(4.9 kg.)**	**Total grains**

6	HBU	(170 MBU) American Perle hops (pellets)—90 minutes (bittering)
¼	oz.	(7 g.) American Tettnanger hops (pellets)— steep in finished boiled wort for 2 to 3 minutes (aroma)

¼	tsp.	Irish moss
¾	c.	corn sugar for priming in bottles. Use ⅓ cup corn sugar if priming a keg.

Wyeast 1968 London ESB Ale yeast

Prior to brewing day, toast oats in a 350 F (177 C) oven by spreading them onto a cookie sheet or screen. Monitor closely every few minutes and turn so that they evenly become dark brown.

A single-step infusion mash is employed to mash the grains. Add 11 quarts (10.5 l.) of 167-degree F (75 C) water to the crushed grain and toasted oats, stir, stabilize and hold the temperature at 150 degrees F (65.5 C) for 60 minutes.

After conversion, raise temperature to 167 degrees F (75 C), lauter and sparge with 4.5 gallons (17 l.) of 170-degree F (77 C) water. Collect about 6 gallons (23 l.) of runoff, add bittering hops and bring to a full and vigorous boil.

The total boil time will be 90 minutes. When 10 minutes remain, add Irish moss. After a total wort boil of 90 minutes (reducing the wort volume to just over 5 gallons), turn off the heat, add aroma hops and let steep 2 to 5 minutes. Then separate or strain out and sparge hops. Chill the wort to 65 to 70 degrees F (18–21 C) and direct into a sanitized fermenter. Aerate the cooled wort well. Add an active yeast culture and ferment for 4 to 6 days in the primary. Then transfer into a secondary fermenter and let age for three more weeks at temperatures of 65 to 70 degrees F (68.5–21 C).

When secondary aging is complete, prime with sugar, bottle or keg. Let condition at temperatures above 60 degrees F (15.5 C) until clear and carbonated.

MASH-EXTRACT RECIPE AND PROCEDURE FOR B&H BREAKFAST TOASTED ALE

2¾	lbs. (1.2 kg.)	English light dried malt extract
2	lbs. (0.9 kg.)	American 2-row Klages pale malt
1¼	lbs. (0.6 kg.)	American wheat malt
1½	lbs. (0.7 kg.)	American Victory malt
1	lb. (0.45 kg.)	oven-toasted flaked oats
½	lb. (0.23 kg.)	English crystal malt—40 Lovibond
¼	lb. (114 g.)	Belgian Biscuit malt
6½	**lbs. (3.0 kg.)**	**Total grains**

7.5	HBU	(213 MBU) American Perle hops (pellets)—60 minutes (bittering)
¼	oz.	(7 g.) American Tettnanger hops (pellets)—steep in finished boiled wort for 2 to 3 minutes (aroma)

¼ tsp.   Irish moss
¾ c.   corn sugar for priming in bottles. Use ⅓ cup
          corn sugar if priming a keg.
Wyeast 1968 London ESB Ale yeast

Prior to brewing day, toast oats in a 350 F (177 C) oven by spreading them onto a cookie sheet or screen. Monitor closely every few minutes and turn so that they evenly become dark brown.

A single-step infusion mash is employed to mash the grains. Add 6 quarts (5.7 l.) of 167-degree F (75 C) water to the crushed grain and toasted oats, stir, stabilize and hold the temperature at 150 degrees F (65.5 C) for 60 minutes.

After conversion, raise temperature to 167 degrees F (75 C), lauter and sparge with 2.5 gallons (9.5 l.) of 170-degree F (77 C) water. Collect about 3 gallons (11.5 l.) of runoff. Add malt extract and bittering hops and bring to a full and vigorous boil.

The total boil time will be 60 minutes. When 10 minutes remain, add Irish moss. After a total wort boil of 60 minutes, turn off the heat, add aroma hops and let steep 2 to 5 minutes. Then separate or strain out and sparge hops, and direct the hot wort into a sanitized fermenter to which 2 gallons (7.6 l.) of cold water have been added. If necessary, add additional cold water to achieve a 5-gallon (19-l.) batch size. Chill the wort to 65 to 70 degrees F (68.5–21 C). Aerate the cooled wort well. Add an active yeast culture and ferment for 4 to 6 days in the primary. Then transfer into a secondary fermenter and let age for three more weeks at temperatures of 65 to 70 degrees F (68.5–21 C).

When secondary aging is complete, prime with sugar, bottle or keg. Let condition at temperatures above 60 degrees F (15.5 C) until clear and carbonated.

Silver Cup Winner

Schierlinger Roggen
Brauerei Thurn & Taxis
Schierling, Germany

Deep amber color with red and orange hues. Notable dense foam and head retention. Bottle-conditioned with yeast and brewed with rye malt. A high, fruity note along with secondary aromas of clove and banana combine to introduce the palate to a beer reminiscent of Bavarian-style wheat beer, though it's not quite the same. A unique spiciness defines the difference. Also evident is a good, sweet, full malt foundation in the aroma. Delightful continuation of aromas into the flavor, with clove and fruitiness quite obvious but not overpowering. The malt foundation continues with no hesitation from the aroma into the flavor. Mouth feel is medium-bodied with a clean, refreshing, well-carbonated aftertaste. Bitterness is quite low but adequate, balancing the full malt foundation. Roggenbier has all the character of a skillful fermentation and is very definitive in what it portrays. There is little confusion on the palate. The malt, spiciness and fruitiness are memorable and enchanting.

Brewery formulation uses German dark, crystal and rye malts with German Perle and Tettnanger hops.

Estimated profile based on tasting
Original Gravity: 1.048 (12) indicated by the brewery
Final Gravity: 1.010 (2.5)
Alcohol by volume: 5% indicated by the brewery
Color: 14–17 SRM (28–34 EBC)
Bittering Units: 14–16 (14 indicated by the brewery)

Bronze Cup Winner

Brewery Hill Honey Amber (Ale)
The Lion Brewery
Wilkes-Barre, Pennsylvania, USA

Amber color with orange hue. Slight herbal, wintergreen hop aroma accompanied by a low-level complex fruitiness. A clean biscuit-cookielike toasted-malt aroma emerges as the beer warms in the glass. Medium-bodied mouth feel is compromised by high carbonation, which tends to create a dry finish. Initial flavor impression suggests toasted and biscuit-aromatic malts. Hop bitterness is portrayed simply with a medium effect, neither harsh nor soft on the palate. The bitterness intensifies in the aftertaste as the sweetness dissipates, leaving bitterness naked and lingering. Pleasant overall balance of hop bitterness, body and malt character. Honey character is not detectable. With no real honey character evident, the sweetness could easily come from the malt.

Estimated profile based on tasting
Color: 8–10 SRM (16–20 EBC)
Bittering Units: 25–29

CATEGORY 60:
SMOKE-FLAVORED BEERS
(ALES OR LAGERS)

A. SUBCATEGORY: BAMBERG-STYLE RAUCHBIER LAGER

Rauchbier should have smoky characters prevalent in the aroma and flavor. The beer is generally toasted-malty-sweet and full-bodied with low to medium hop bitterness. Noble hop flavor is low but perceptible. Low noble hop aroma is optional. The aroma should strike a balance between malt, hop and smoke. Fruity esters, diacetyl and chill haze should not be perceived.

Original Gravity (°Plato):	1.048–1.052 (12–13 °Plato)
Apparent Extract–	
Final Gravity (°Plato):	1.012–1.016 (3–4 °Plato)
Alcohol by weight (volume):	6–4% (4.3–4.8%)
Bitterness (IBU):	20–30
Color (SRM):	10–20 (20–80 EBC)

B. SUBCATEGORY: SMOKED-FLAVORED BEER (LAGER OR ALE)

Any style of beer can be smoked; the goal is to reach a balance between the style's character and the smoky properties.

Gold Cup Winner

Aecht Schlenkerla Rauchbier
Brauerei Heller-Trum
Brauereiausschank Schlenkerla
Dominikanerstraβe 6
96049 Bamberg, Germany
Brewmaster: German Trum
Established 1678
Production: 42,700 bbl. (50,000 hl.)

Tucked away in the heart of the Franconia region of Germany north of Munich, the Heller-Trum Brewery with its popular brewery restaurant Schlenkerla is a German trea-

German Trum

Inside the brewery

sure. Franconia is home to more breweries per capita than all the rest of Germany. Bamberg, a small city of about 80,000, has nine breweries within the city and 100 within a short drive. Aecht Schlenkerla Rauchbier is made in the Märzen lager style from 100 percent smoked malt, and mashed using the traditional German double-decoction method. The malt is smoked with beech wood, bottom-fermented and matured for seven weeks. The tradition of smoked beers predates all the modern methods of making beer today.

Hundreds of years ago all beers likely were Rauchbier, German for "smoked beer." Before modern malt drying techniques were developed, barley would be soaked in water, sprouted and then dried over the heat of wood fires. The smoke character that was imparted became a signature character of medieval beers. This ancient tradition was purposefully preserved by the Heller brewery and several others in Bamberg. The name Aecht Schlenkerla roughly translated from old German means "original flailer." It was said that the brewer who formulated the recipe used to walk, cane in hand, in a wildly exaggerated swagger, with arms flailing. Look closely at the label and you will see him there. Drink this beer with gusto and you, too, may swagger with a smile.

CHARACTER DESCRIPTION OF GOLD CUP–WINNING AECHT
SCHLENKERLA RAUCHBIER

The warmth of the fire from which this beer sprang is reflected in its deep, tawny brown color with hints of orange and red. If you enjoy the aroma of gently sweetened smoke, Aecht Schlenkerla Rauchbier is heaven on earth, soft and comforting like a friendly fire. Spending some quality time with a glass of Schlenkerla, you will note there's something special behind the smoke. Like honey, the toasted malt character comes through with a special sweetness. The quality of the beech-wood smoke and the softness of its impression evoke great memories of campfires and fireside friendships. That said, note that I haven't even begun to discuss the beer's flavor. First flavor impression is briefly smoke, followed by the impact of the toasted sweet malt character. The complex synergy and balance of malt honey sweetness, noble-hop-flavored bitterness and smoke immediately follow, completing the sojourn. Though only 100 percent smoked malt is used, Schlenkerla has a Munich malt character: full to medium in body and mouth feel with no caramel, toffee, biscuit or toasted character whatsoever. Interestingly, this beer is provocatively smoky in aroma and flavor, but the aftertaste and impression is one of trailing balanced bitterness and a surprisingly clean palate.

• *Recipe for 5 U.S. gallons (19 liters) Aecht Schlenkerla Rauchbier*
  Targets:
  Original Gravity: 1.054 (13.5)
  Final Gravity: 1.014 (3.5)

Alcohol by volume: 5%
Color: 15 SRM (30 EBC)
Bittering Units: 30

ALL-GRAIN RECIPE AND PROCEDURE

10   lbs. (4.5 kg.)   German/Bamberg Rauch (smoked)
                        malt—10 Lovibond
10   lbs. (4.5 kg.)   Total grains

6   HBU (170 MBU) German Hersbrucker Hallertauer
                    hops (pellets)—75 minutes (bittering)
1.5 HBU (43 MBU) German Hersbrucker Hallertauer
                   hops (pellets)—30 minutes (flavor)

¼   tsp. Irish moss
¾   c.   corn sugar for priming in bottles. Use ⅓ cup
           corn sugar if priming a keg.
Wyeast 2206 Bavarian Lager yeast

A double-decoction mash is employed to mash the grains.
If you are using alkaline or hard water, treat the water with
an appropriate amount of lactic acid or refer to pages 120
to 122 of *The Home Brewer's Companion* for proceeding with a
triple decoction with an acid rest.

Add 10 quarts (9.5 l.) of boiling water to the grain, raising
the temperature to 122 degrees F (50 C). Then remove 5
quarts (4.8 l.) of the thickest mash and boil this in another
vessel for 30 minutes. Stir this boiled decoction constantly.
Maintain the remainder of the mash at 122 degrees F (50
C) while boiling the decocted portion. After boiling for 30
minutes, add the decoction back into the main mash vessel,
raising the temperature to 150 degrees F (65.5). Maintain
this temperature for one hour. Up to 10 quarts (9.5 l.) of
boiling water may be added at any time during this period
to maintain mashing temperature. After one hour, remove
one third of the mash (a fifty-fifty blend of thick mash and
liquid) and boil about 20 minutes. Stir this second decoc-

tion constantly during the boil. When finished, return the boiled decoction to the main mash vessel, ending starch conversion by having raised the temperature to about 167 degrees F (75 C). Then lauter and sparge with 4 gallons (15 l.) of 170-degree F (77 C) water. Collect about 6 gallons (23 l.) of runoff, add bittering hops and bring to a full and vigorous boil.

The total boil time will be 90 minutes. When 30 minutes remain, add flavor hops. When 10 remain, add Irish moss. After a total wort boil of 90 minutes (reducing the wort volume to just over 5 gallons), turn off the heat, then separate or strain out and sparge hops. Chill the wort to 60 degrees F (15.5 C) and direct into a sanitized fermenter. Aerate the cooled wort well. Add an active yeast culture and ferment for 4 to 6 days in the primary. Then transfer into a secondary fermenter, chill to 50 degrees F (10 C) to age for two more weeks, then lager for four more weeks at 40 degrees F (4.5 C).

When secondary aging is complete, prime with sugar, bottle or keg. Let condition at temperatures above 60 degrees F (15.5 C) until clear and carbonated, then store chilled.

MASH-EXTRACT RECIPE AND PROCEDURE FOR AECHT SCHLENKERLA RAUCHBIER

2½	lbs. (1.3 kg.)	English light dried malt extract
6	lbs. (2.7 kg.)	German/Bamberg Rauch (smoked) malt—10 Lovibond
6	**lbs. (2.7 kg.)**	**Total grains**

7	HBU	(198 MBU) German Hersbrucker Hallertauer hops (pellets)—60 minutes (bittering)
1.5	HBU	(43 MBU) German Hersbrucker Hallertauer hops (pellets)—30 minutes (flavor)

¼	tsp.	Irish moss
¾	c.	corn sugar for priming in bottles. Use ⅓ cup corn sugar if priming a keg.

Wyeast 2206 Bavarian Lager yeast

This version will not be as full in smoke flavor as the all-grain version, but will still provide a memorable impression.

A double-decoction mash is employed to mash the grains. If you are using alkaline or hard water, treat the water with an appropriate amount of lactic acid or refer to pages 120 to 122 of *The Home Brewer's Companion* for proceeding with a triple decoction with an acid rest.

Add 6 quarts (9.5 l.) of boiling water to the grain, raising the temperature to 122 degrees F (50 C). Then remove 3 quarts (1.4 l.) of the thickest mash and boil this in another vessel for 30 minutes. Stir this boiled decoction constantly. Maintain the remainder of the mash at 122 degrees F (50 C) while boiling the decocted portion. After boiling for 30 minutes, add the decoction back into the main mash vessel, raising the temperature to 150 degrees F (65.5). Maintain this temperature for one hour. Up to 6 quarts (8.6 l.) of boiling water may be added at any time during this period to maintain mashing temperature. After one hour, remove one third of the mash (a fifty-fifty blend of thick mash and liquid) and boil about 20 minutes. Stir this second decoction constantly during the boil. When finished, return the boiled decoction to the main mash vessel, ending starch conversion by having raised the temperature to about 167 degrees F (75 C). Then lauter and sparge with 2 gallons (7.6 l.) of 170-degree F (77 C) water. Collect about 3 gallons (11.4 l.) of runoff, add bittering hops and bring to a full and vigorous boil.

The total boil time will be 60 minutes. When 30 minutes remain, add flavor hops. When 10 remain, add Irish moss. After a total wort boil of 60 minutes, turn off the heat, then separate or strain out and sparge hops, and direct the hot wort into a sanitized fermenter to which 2 gallons (7.6 l.) of cold water have been added. If necessary, add additional cold water to achieve a 5-gallon (19-l.) batch size. Chill the wort to 60 degrees F (15.5 C). Aerate the cooled wort well. Add an active yeast culture and ferment for 4 to 6 days in the primary. Then transfer into a secondary fermenter, chill to 50 degrees F (10 C) to age for two more weeks, then lager for four more weeks at 40 degrees F (4.5 C).

When secondary aging is complete, prime with sugar, bottle or keg. Let condition at temperatures above 60 degrees F (15.5 C) until clear and carbonated, then store chilled.

Silver Cup Winner

Alaskan Seasonal Smoked Porter
Alaskan Brewing & Bottling Co.
Juneau, Alaska, USA

Deep, dark brown color with hints of laser red emerging from around the edges. High impact of smoke aroma is further accented by the lack of malt sweetness in aroma. Mouth feel is like that of a medium-bodied porter. Sweet caramel-malt character is evident in flavor and aftertaste. Roast-malt flavor comes through as well. Alaskan alder smoked flavor is well balanced with malt sweetness, roast malt character and hop bitterness. A difficult balance to achieve as both roast malt and smoked malt can lend their own bitterness. The aftertaste is clean with only a comparatively mild and lingering smoke character. Smoked Porter is a unique addition to the world of classics. This is an achievement in itself. Overall impression is smoke flavor accompanied by a robust bitterness followed by a faint memory of sweet caramel/malt. Cleanly fermented.

Estimated profile based on tasting
Color: 24+SRM (48+EBC)
Bittering Units: 40–45

Bronze Cup Winner

None

CATEGORY 61:
NONALCOHOLIC MALT BEVERAGES

Nonalcoholic (NA) malt beverages should emulate the character of a previously listed category/subcategory designation but without the alcohol (less than 0.5 percent).

Gold Cup Winner

Radegast Birell
Radegast Brewery J.S.C.
739 51 Nošovice, Czech Republic
Brewmaster: Ing. Stanislav Fridrich
Established 1970
Production: 171,000 bbl. (200,000 hl.)

Besides its award-winning nonalcoholic Radegast Birell, the brewery produces Radegast Premium, Radegast Triumf and Radegast Light.

CHARACTER DESCRIPTION OF GOLD CUP–WINNING RADEGAST BIRELL
Golden yellow and extremely pale. Aroma is honestly reminiscent of real beer, though understandably slightly worty and malty. Aroma is sweet with a very nice noble German or herbal

Ing. Stanislav Fridrich

Czech Saaz hop character. A hint of fermentation character sneaks through, but the aroma of fresh, wonderful wort predominates. First flavor impression indicates a very light bodied beer with watery mouth feel. Bitterness lingers very refreshingly with gentle aftertaste of hops and malt extract, which is refreshing with such a light body. Hop aroma continues to be recalled while indulging. Flavor is balanced toward a soft hop bitterness and excellent hop flavor. The art of balancing hop flavor, aroma and bitterness is essential in formulating the recipe of this low-gravity "beer."

- *Recipe for 5 U.S. gallons (19 liters) Radegast Birell*
  Targets:
  Original Gravity: 1.022 (5.5)
  Final Gravity: 1.018 (4.5)
  Alcohol by volume: 0.5%
  Color: 4 SRM (8 EBC)
  Bittering Units: 24

ALL-GRAIN RECIPE AND PROCEDURE

4 lbs. (1.8 kg.) German Pilsener malt
4 lbs. (1.8 kg.) Total grains

4 HBU (113 MBU) German Hersbrucker Hallertauer hops (pellets)—60 minutes (bittering)
2.5 HBU (71 MBU) Czech Saaz hops (pellets)—30 minutes (flavor)
¼ oz. (7 g.) Czech Saaz hops (pellets)—steep in finished boiled wort for 2 to 3 minutes (aroma)

¼ tsp. Irish moss
Special yeast for producing nonalcoholic beers not commercially available to the homebrewer

The process of producing nonalcoholic (less than 0.5 percent alcohol by volume) "beer" is a complex one that requires special yeasts and equipment. Radegast Birell is probably made using special yeast and yeast management. The wort is overoxygenated and chilled to temperatures below 40 degrees F (4.5 C), then a special yeast is introduced. It is forced to enter and remain in the respiration cycle (see page 371, *The Home Brewer's Companion*), where it metabolizes some sugar into carbon dioxide, some flavor compounds and very little alcohol. After this phase is completed, the yeast is removed by filtration and the beer is carefully maintained in sterile conditions before bottling and pasteurization. What follows is a likely scenario for wort formulation. Homebrewed fermentation techniques are not discussed due to the impracticality of the process for homebrewers.

A double-decoction mash is employed to mash the grains. If you are using alkaline or hard water, treat the water with an appropriate amount of lactic acid or refer to pages 120 to 122 of *The Home Brewer's Companion* for proceeding with a triple decoction with an acid rest.

Add 4 quarts (3.8 l.) of boiling water to the grain, raising the temperature to 122 degrees F (50 C). Then remove 2 quarts (1.9 l.) of the thickest mash and boil this in another vessel for 30 minutes. Stir this boiled decoction constantly. Maintain the remainder of the mash at 122 degrees F (50 C) while boiling the decocted portion. After boiling for 30 minutes, add the decoction back into the main mash vessel, raising the temperature to 150 degrees F (65.5). Maintain this temperature for one hour. Up to 4 quarts (3.8 l.) of boiling water may be added at any time during this period to maintain mashing temperature. After one hour, remove one third of the mash (a fifty-fifty blend of thick mash and liquid) and boil about 20 minutes. Stir this second decoction constantly during the boil. When finished, return the boiled decoction to the main mash vessel, ending starch conversion by having raised the temperature to about 167 degrees F (75 C). Then lauter and sparge with 2 gallons (7.6

l.) of 170-degree F (77 C) water. Collect about 3 gallons (8.6 l.) of runoff and add more water to attain a total volume of 5.5 gallons (22 l.). Add bittering hops and bring to a full and vigorous boil.

The total boil time will be 60 minutes. When 30 minutes remain, add flavor hops. When 10 minutes remain, add Irish moss. After a total wort boil of 60 minutes (reducing the wort volume to just over 5 gallons), turn off the heat and add aroma hops, letting steep for 2 to 3 minutes. Then separate or strain out and sparge hops. Chill the wort to below 40 degrees F (4.5 C) and ferment using nonalcoholic beer "fermentation," filtration and pasteurization techniques.

When secondary aging is complete, force-carbonate, counterpressure bottle and pasteurize.

MASH-EXTRACT RECIPE AND PROCEDURE FOR RADEGAST BIRELL

2½	lbs. (1.1 kg.)	English light dried malt extract
0	**lb. (0 kg.)**	**Total grains**

4	HBU	(113 MBU) German Hersbrucker Hallertauer hops (pellets)—60 minutes (bittering)
2.5	HBU	(71 MBU) Czech Saaz hops (pellets)—30 minutes (flavor)
¼	oz.	(7 g.) Czech Saaz hops (pellets)—steep in finished boiled wort for 2 to 3 minutes (aroma)

¼	tsp.	Irish moss

Special yeast for producing nonalcoholic beers not commercially available to the homebrewer

See all-grain recipe for introduction to process.

Add bittering hops and malt extract to 2.5 gallons (9.5 l.) of water and bring to a full and vigorous boil. The total boil time will be 60 minutes. When 30 minutes remain, add flavor hops. When 10 minutes remain, add Irish moss. After a total wort boil of 60 minutes, turn off the heat and add aroma hops, letting steep for 2 to 3 minutes. Then separate or strain out and sparge hops. Chill the wort to below 40 degrees F (4.5 C) and direct into a sanitized fermenter. Aerate the cooled wort well and ferment using nonalcoholic beer "fermentation," filtration and pasteurization techniques.

When secondary aging is complete, force-carbonate, counterpressure bottle and pasteurize.

Silver Cup Winner

None

Bronze Cup Winner

None

Association of Brewers
1996 World Beer Cup

Staff

Marcia Schirmer, Director
Jeanne Colon-Bonet, Judging Manager
Glenn Colon-Bonet, Staging Manager
Sharon Mowry, Beer Service Manager
Sheri Winter, Marketing Director
Charlie Papazian, President
    Association of Brewers
Cathy B. Ewing, Vice President
    Association of Brewers
Bob Pease, Beer Service
    Association of Brewers
Kyle Keazer, Beer Service
    Association of Brewers
Melinda Bywaters, Marketing
    Association of Brewers
Rebecca Bradford, Staging
    Association of Brewers

Competition Captains
and Stewards

Ron Moucka, Assistant Manager
John Adams
John Bates
Kris Bennett
John Paul Broad
John Carlson
Lynn Danielson
Will Darlington
Nanci Dexter
Chuck Fowler
Ann Glasier
Jason Goldman
Scott Mills
Gretchen Neudeck
Keith Schwols

Boyd Smart
Mara Sprain
Howard Stroyan
Mike Stroyan
Brian Walter
Chris Wright

1996 World Beer Cup Judges

Jeff Alexander, former Head Brewer
Los Gatos Brewing Co.
Los Gatos, California, USA

Fal Allen, Head Brewer
Pike Brewing Company
Seattle, Washington, USA

Paul Bayley, Head Brewer
Marstons, Thompson & Evershed
Staffs, England

David Bruce, Founder
Belcher's Brewery
Berkshire, England

Fred Eckhardt
Fred Eckhardt Communications
Portland, Oregon, USA

Teri Fahrendorf, Head Brewer
Steelhead Brewery & Cafe
Eugene, Oregon, USA

Pat Goddard, Fellow
Institute of Brewing, Toohey's Ltd.
Lindcombe, New South Wales, Australia

John Harris, Brewmaster
Full Sail Brewing Co.
Hood River, Oregon, USA

Ray Klimovitz, Director, Brewing Development
The Stroh Brewery Co.
Detroit, Michigan, USA

Grant Johnston, Head Brewer
Third Street Ale Works
Santa Rosa, California, USA

Finn Knudsen, President
Beverage Consult International
Evergreen, Colorado, USA

Brad Kraus, Head Brewer
Wolf Canyon Brewing Co.
Santa Fe, New Mexico, USA

Ed Laperle, Analyst
Ball Corporation
Muncie, Indiana, USA

Geoff Larson, Brewmaster
Alaskan Brewing & Bottling Co.
Juneau, Alaska, USA

Dick Leach, Corporate Brewer
Molson Breweries
Etobicoke, Ontario, Canada

Jeff Lebesch, Head Brewer
New Belgium Brewing Co.
Fort Collins, Colorado, USA

Gary Luther, Senior Staff Brewer
Miller Brewing Company
Milwaukee, Wisconsin, USA

John Mallett
Saaz Brewing Equipment & Services
Arlington, Virginia, USA

Ryouji Oda, President
Japan Craft Beer Association
Ashiya City, Japan

Brad Page, Brewmaster
CooperSmith's Pub and Brewery
Fort Collins, Colorado, USA

Charlie Papazian, President
Association of Brewers
Boulder, Colorado, USA

Tom Schmidt, Director, Beverage Development, Brewing
Anheuser-Busch Inc.
St. Louis, Missouri, USA

David Sipes, Head Brewer
Sudwerk Privatbrauerei Hubsch
Davis, California, USA

Hube Smith, former Head Brewer
Wild River Brewing Company
Cave Junction, Oregon, USA

Wayne Waananen, former Head Brewer
The Sandlot Brewery
Denver, Colorado, USA

Eric Warner, Brewmaster
Tabernash Brewing Co.
Denver, Colorado, USA

Marty Watz, Resident Brewmaster
Anheuser-Busch, Inc.
Fort Collins, Colorado, USA

1996 World Beer Cup Sponsors

Anheuser-Busch, Inc.
Fort Collins, Colorado
Host Distributor

Anheuser-Busch, Inc.
St. Louis, Missouri
Importer of Record

Everbrite
Greenfield, Wisconsin

Judge Sponsors

Haertling Custom Awards
Boulder, Colorado
Design

Sahm GmbH & Co. KG
Germany
Glassware

Sonnenalp Resort
Vail, Colorado
Hospitality

Tompkins World Travel Services
Mill Valley, California
Travel Services

Westway Express
Denver, Colorado
Transportation

Index

417